CALCULUS

CALCULUS

Second Edition

Edward G. Begle
Stanford University

Lloyd B. Williams
Reed College

HOLT, RINEHART AND WINSTON, INC.
New York Chicago San Francisco Atlanta
Dallas Montreal Toronto London Sydney

This book was originally published in 1954 as *Introductory Calculus with Analytic Geometry* by Edward G. Begle.

Library of Congress Catalog Card Number: 73-159399
ISBN: 0-03-082800-7
AMS 1970 Subject Classification Numbers: 26-01, 26A06, 26A09, 98A20
Printed in the United States of America
9 8 7 6 5 4 3 2 1 038 5 4 3 2

PREFACE

Professor Edward G. Begle wrote *Introductory Calculus with Analytic Geometry* in 1954. It was quickly recognized as an effective exposition of elementary calculus presented as a sample of a mathematical discipline with emphasis on its essentially mathematical character. He pointed out applications of calculus to other scientific fields, but he treated the subject with attention to its own content and consistency. Over the years, we have found that this treatment appeals to students who are inclined to study the subject for its own sake.

With this experience, we have rewritten the book. In doing so, we have adhered to the aims expressed by Professor Begle in the Preface to the first edition.

> Our aim in presenting calculus in this fashion is to give the student more of an understanding of the basic concepts of the subject than is usually done in an introductory course. It is our belief that such an understanding is becoming more and more necessary for the student of science or engineering. In these fields, the mathematical techniques employed have become so numerous and varied that no student can be expected to master more than a few of them without a framework of theory around which they can be organized.
>
> We also feel that for the student for whom this is a terminal course in mathematics it is desirable to show the logical structure of calculus rather than to concentrate entirely on the problem-solving techniques which are the end products of calculus.

The teaching of mathematics in secondary schools has improved over the last decade and a half; college students are now better prepared to undertake a college calculus course. In our new version of the book we have responded to this improved preparation by reducing the attention previously given by Professor Begle to certain topics that are now commonly taught in good high school courses. In particular, we presume the students are familiar with mathematical induction, computational algebra, set notation, rudimentary analytic geometry (for graphing), geometrical trigonometry, and some logic used in proofs.

This book is meant to be used with a teacher in conventional class sessions; it is not programmed in the narrow sense, nor is it meant to answer all the student's questions as he reads it. Rather, it should generate many topics for class discussion. Very likely certain of the proofs will be revised in class, and this is expected to be part of the exercise. We do claim to present, briefly, the essential theorems of elementary calculus, and to provide motivation for their introduction. The techniques and definitions are chosen carefully so as to be representative of analysis, and to be useful and expandable if the student goes on to more advanced work.

The treatment frankly is axiomatic-deductive; the axioms being those for the real number system set down in Chapter 1. Although the real number system is not treated in detail, many of its properties are cited, and special emphasis is put on the least upper bound axiom. Functions and the calculus are built by definitions applied to the real number system. We define a function to be a domain with an associated rule. The set of ordered pairs of numbers, which we call the "graph" of the function, is used in places where it is helpful. We feel that the domain and rule is close to the way we think about a function, and that it leads a student toward the concept of a mapping. We expect the student will eventually conceive of functions as mathematical objects. He must reach this threshold before endeavoring to apply analytic techniques to them.

Chapter 3 is an exercise in these analytic techniques, motivated by the idea of unbrokenness of the geometrical graph, which is a "plot" of the function. Conventional ϵ-δ procedures first appear here and these are used to delineate continuous functions. There is a standard treatment of differentiation in Chapter 4 based on a study of smoothness of the geometrical graph. We might apologize for the symbol, ${}_f\sigma_x$, the slope function for the function f at an accumulation point x of its domain, but this symbol does serve initially to introduce the difference quotient as, in fact, a function. The symbol is effectively abandoned after Chapter 4. In Chapter 5, we treat the extreme value problem, and we use this topic to illustrate an important use of the mean value theorem. To us, the slope of a tangent line is the canonical interpretation of the value of a derivative; that is, we tend to think geometrically. However, we emphasize that the derivative has many other useful interpretations as rates.

The Riemann-Darboux integral of a continuous function is the subject of Chapter 6, and the treatment is conventional. Here too, although the geometric interpretation as an area is central, we treat other models for the integral. The section on the Riemann integral is the only place at which we need uniform continuity.

Chapter 7, on elementary functions, applies most of the preceding analytic theory to the construction of log and arctan. It is likely that the student may feel he knows these functions already, and he may hesitate to recapture them in such a meticulous way. Therefore, we help the teacher lead the class into this project by posing the problem of practical integration. We conclude with a minimum of techniques of integration in Chapter 8. Students usually like to "do" integrations so we end the course on a relaxed note.

Given the preparation we assume students bring to the course, the Appendixes are not essential. However, they should generate discussion and can be used for special class projects.

This book records an intensive course; it is written for students with a somewhat longer than average attention span. It should increase the students' ability to handle mathematical abstraction by means of written symbolism and to keep the two properly in their separate roles. The sample of mathematics we choose is relatively brief, neat, precise, and can be treated with a logic which is really just refined common sense.

Palo Alto, California — Edward G. Begle

Portland, Oregon — Lloyd B. Williams

October 1971

CONTENTS

CALCULUS

INTRODUCTION

Mathematics is a particular body of knowledge. It has its own language which combines the language of everyday use (English in this book) with a special mathematical language having a vocabulary and grammar peculiar to it. The vocabulary of mathematics contains and uses a mixture of English and technical symbols, including "circle," "triangle," "2," "number," "+," "$\sqrt{}$," "function," and many others. The grammar tells us, for example, that the sentence $a + b = \sqrt{a - b}$ is grammatically correct, while the sentence $a^+ - b = \sqrt{\quad}$ is not. The literature of mathematics contains theorems together with their proofs. This literature is important and interesting. Furthermore, it is well suited to the discussion of exact, quantitative concepts, and is indispensable in the daily work of scientists, engineers, and many other professional people.

This book is one chapter in the literature of mathematics, a chapter entitled *Calculus*. We will describe this subject, and we will indicate from time to time how calculus is related to that world of our physical senses with which physical science deals.

Compared with methods used in other disciplines, the method employed in mathematics is unique in one respect. The logically fundamental statements of a mathematical theory are few, and they can be written down explicitly at the beginning of the development of the theory. Although for the sake of clarity some of these fundamental statements are postponed until later in the discourse, they can be listed at the outset and all theorems of the theory can be derived from them by a sequence of logical steps. The fundamental statements are called *axioms*.

The student is already familiar with axioms and theorems for geometry. He may have seen an axiomatic treatment of algebra. Chapter 1 of this book contains axioms for the system of real numbers. The purpose is twofold. First, in this way we can present a sample of the uniquely mathematical method of reasoning and we can become familiar with techniques that are useful in calculus. Second, we can describe a logically sound foundation for the number sets which are to be the domains and ranges of the functions to be studied.

Beginning with Chapter 2 we deal specifically with calculus and some of its uses. Throughout, it is assumed that the student knows certain ground rules, such as the logic used in proofs and certain terminology and symbols. Occasionally we call attention to such rules, but it is not necessary here, and certainly beyond the scope of this book, to develop a complete "metamathematical" foundation for calculus. Some brief comments suffice for a start.

Let us agree on what *equality* means. We consider this to precede our theory with its axioms. In Chapter 1 lowercase letters stand for real numbers. Which number a particular letter stands for is determined by the conditioning statements in the local context. If "x is a real number" is asserted alone, then one may choose any real number for x. If "x is a real number whose square is 9" is stated, then x stands only for 3 or for -3. The statement "$x = y$" means that x and y must stand for the same number. With this, it follows immediately that "if $x = y$, then $y = x$" is true; and so, also, is "if $x = y$ and $y = z$, then $x = z$." Generally, unless a specific warning is given, the meaning of a letter is not changed during a particular discussion. This convention was assumed in the sentence before last. In this light, "$x = x$" is universally true.

An *operation* in a set of objects (such as a set of numbers, a set of ordered pairs of numbers) is a rule that assigns to each of certain ordered pairs of these objects a unique object of the set. For example, in the set of whole numbers, the addition operation assigns to the ordered pair 2,3 the number 5. This assigned number, 5, is designated by $2 + 3$, and since both 5 and $2 + 3$ stand for this unique sum, it is true that $2 + 3 = 5$. Division is an operation in the set of whole numbers also, but some ordered pairs of numbers are left out; for example, $2 \div 3$ does not represent a whole number. Different operations may have different properties. For instance, in the set of integers (positive and negative whole numbers with zero) addition is commutative ($a + b = b + a$ for every a and b), but subtraction is not ($a - b \neq b - a$ for most choices of a and b).

The problems in the Introduction are to help the student start to think about mathematics again. The systematic development of calculus of our theory starts in Chapter 1.

PROBLEMS

1. Each of the equations below is a statement about integers.
   (a) Is each statement in the sequence
   $$x = 4, \quad x^2 = 16, \quad x = 4 \quad \text{or } x = -4$$
   a logical consequence of its predecessor?
   (b) Ask the question in (a) for
   $$x = 4 \quad \text{or} \quad x = -4, \quad x^2 = 16, \quad x = 4.$$
   (c) Ask the same question in (a) for
   $$\sqrt{x+6} = 1 - \sqrt{x-5}\,, \quad x + 6 = x - 4 - 2\sqrt{x-5}\,,$$
   $$5 = -\sqrt{x-5}\,, \quad 25 = x - 5, \quad x = 30.$$
   (d) Why is there no real number that x can stand for so that $\sqrt{x+6} = 1 - \sqrt{x-5}$?
2. In the system of integers (positive and negative whole numbers with zero) the following theorem is true.

   THEOREM Given $m > 0$, then for any integer x there is exactly one pair of integers q and r, so that
   $$x = mq + r, \quad 0 \leqslant r < m.$$
   (a) Let $m = 3$. Find q and r for each of $x = 0, 1, 2, 3, 4, 5, 6, 7$, and the negatives of these.
   (b) With $m = 3$, find q and r for each of $x = -37, -20, 37$, and 20.
   (c) For a fixed $m > 0$, let us say r represents x if $x = mq + r$, $0 \leqslant r < m$, for some q.

   DEFINITION $x \cong y\ (m)$ in case x and y are represented by the same integer r. (Read x "congruent" to y modulo m.) Prove
   (a) $x \cong x$ for all integers x;
   (b) if $x \cong y$, then $y \cong x$;
   (c) if $x \cong y$ and $y \cong z$, then $x \cong z$.

Notice the similarity between $\cong$ and $=$.

3. In the system of rational numbers, suppose that we tried to define a new operation by
   (a) $\frac{a}{b} \oplus \frac{c}{d} = \frac{a+c}{b+d}$. Would $\oplus$ be commutative? Associative?
   (b) $\frac{a}{b} \otimes \frac{c}{d} = \frac{ad}{bc}$. Would $\otimes$ be commutative? Associative?
4. In the system of integers, two numbers are equal only if they look the same; that is, $5 = 5$ but $5 \neq 7$. However, in the rational number system two numbers can be equal even though they look different; for example, $6889/332 = 83/4$. How do you explain the distinction in usage of $=$?

Chapter 1 REAL NUMBERS AND THEIR ARITHMETIC

1.1 AXIOMS

We will put down a set of axioms for the real number system. Consider a set containing at least two elements in which there are two operations, *addition,* symbolized by $+$, and *multiplication,* symbolized by $\cdot$. If a and b stand for elements of this set, then there is a unique element of the set assigned to the ordered pair a,b by the operation $+$, and $a+b$ stands for this element. Notice that a and b may stand for the same number. Similarly, there is a unique element $a \cdot b$ assigned to the ordered pair a,b by the operation $\cdot$. Since we have agreed upon the meanings of equality and operation, it necessarily is true that if $a = b$, then both of $a + c = b + c$ and $ac = bc$ are true for every c (the abbreviation ab is used for $a \cdot b$). Let us refer to elements of the given set as *numbers,* to be endowed with certain properties by the axioms.

Axioms for addition follow.

A1 If a and b are numbers, then $a + b = b + a$.

A2 If a, b, and c are numbers, then $(a + b) + c = a + (b + c)$.

A3 There is a number[1] 0 (called *zero*) such that for each number a, $a + 0 = a$.

A4 To each number a there corresponds a number $-a$ (called the *negative* of a) such that $a + (-a) = 0$.

Axioms for multiplication follow.

M1 If a and b are numbers, then $ab = ba$.

M2 If a, b, and c are numbers, then $(ab)c = a(bc)$.

M3 There is a number[2] 1 (called *one*) such that for each number a, $a \cdot 1 = a$.

M4 To each number a, except 0, there corresponds a number 1/a (called the *reciprocal* of a) such that $a(1/a) = 1$.

The distributive axiom follows.

D If a, b, and c are numbers, then $a(b + c) = ab + ac$.[3]

Before completing the list of axioms for the real number system, we pause to observe some immediate consequences of the axioms just written. If we choose a zero, then $0 + a = a$ for every a by Axiom A1, and $(-a) + a = 0$. (Contrast these with Axioms A3 and A4.) Likewise, $1 \cdot a = a$ and $(1/a)a = 1$ for any chosen "one." (See Axioms M3 and M4.) Although Axiom A3 leaves open the possible existence of several zeros, the conjunction of Axioms A1 and A3 implies that the zero is unique, as the following theorem demonstrates.

THEOREM 1.1 If 0 and $0'$ each stand for a zero of Axiom A3, then $0 = 0'$.

Proof:

$0 + 0' = 0$	A3 using $0'$ for the zero
$0' + 0 = 0'$	A3 using 0 for the zero
$0 + 0' = 0' + 0$	A1
$0 = 0'$.	first two steps with the ground rule for equality

Thus, 0 and $0'$ necessarily stand for the same element of the set of numbers. It is impossible to name two elements as zeros and keep

[1] According to Axiom A3, there may be more than one such number. But see Theorem 1.1, whose proof shows how Axiom A1 precludes this.

[2] See Footnote 1 and Theorem 1.2.

[3] $ab + ac$ means $(ab) + (ac)$.

Axioms A1 and A3 as well as our ground rules about equality and operations.[4] ■

The choice of a "one" is easy also, for there is only one of these.

THEOREM 1.2 If 1 and $1'$ each stand for a "one" of Axiom M3, then $1 = 1'$.

Proof: The proof is similar to the proof of Theorem 1.1. ■

However, the zero and the one must be different numbers. To see why, we go deeper into the structure of the number system created by the axioms. To begin with, there might be a number $0'$ such that $a + 0' = a$ for some particular a, but still this $0'$ is not the zero of Axiom A3, which has this property for all a. We settle the question with the next theorem.

THEOREM 1.3 If $0'$ is a number such that $a + 0' = a$ for some a, then $0' = 0$.

Proof:

$a + 0' = a$	given for some a
$(-a) + (a + 0') = (-a) + a$	Use a $-a$ given by A4.
$((-a) + a) + 0' = 0$	A2, A4
$0 + 0' = 0$	A1, A4
$0' = 0.$	A1, A3 ■

Notice that no reference was made to the definition of an operation in the reason for the second step, where it should be used. We will gradually condense the proofs of theorems as you learn to supply the omitted steps and their reasons. Be sure that omitted details are clear before you read on.

In the A axioms zero is associated with +, and in the M axioms one is associated with $\cdot$. Axiom D combines + with $\cdot$, and by means of this combination we get a multiplicative property of zero.

THEOREM 1.4 If a is a number, then $a \cdot 0 = 0$.

Proof:

$a \cdot 0 + a \cdot 0 = a(0 + 0)$	D
$a(0 + 0) = a \cdot 0$	A3
$a \cdot 0 + a \cdot 0 = a \cdot 0$	definition of equality
$a \cdot 0 = 0.$	In Theorem 1.3 take $a \cdot 0$ for a and $a \cdot 0$ for $0'$ ■

Now we have a bridge connecting the zero (for +) and the one (for $\cdot$) which we can use to prove the next theorem.

[4] We indicate the end of a proof by the symbol ■

THEOREM 1.5 $1 \neq 0$.

Proof: Assume that $1 = 0$. Because there is more than one number in the set of numbers, we can choose another, say a. Then

$a \cdot 1 = a \cdot 0$	definition of operation
$a \cdot 1 = a$	M3
$a \cdot 0 = 0$	Theorem 1.4
$a = 0.$	definition of equality

Therefore, if $1 = 0$, then every number is equal to the unique zero. The set we are studying could then have only one element in it, contrary to our agreement that it is to have at least two. Hence, $1 \neq 0$. ■

1.1 PROBLEMS

1. Prove that $0 + a = a$ for every number a.

   Answer:

$a + 0 = a$	A3
$a + 0 = 0 + a$	A1
$0 + a = a.$	definition of equality

2. Prove that $1 \cdot a = a$ for every number a.
3. Prove that, given $a \neq 0$, then $(1/a)a = 1$.

   Answer:

$a(1/a) = 1$	M3
$a(1/a) = (1/a)a$	M1
$(1/a)a = 1.$	definition of equality

4. Prove that $(-a) + a = 0$ for every number a.
5. Prove Theorem 1.2.

   Answer: Follow the pattern of the proof of Theorem 1.1.

6. Prove that $-a = (-1)a$ for every number a.

   Answer:

There is a number 1.	M3, Theorem 1.2
There is a number -1.	M4, Theorem 1.6, (Section 1.2)
$1 \cdot a = a$	problem 2
$a + (-1)a = 1 \cdot a + (-1)a$	definitions of operation and equality
$1 \cdot a + (-1)a = (1 + (-1))a$	D, M1
$1 + (-1) = 0$	A3
$a + (-1)a = 0 \cdot a$	definition of equality
$a + (-1)a = 0$	Theorem 1.4
$(-1)a = -a.$	M4, Theorem 1.6 (Section 1.2)

   Note: Check the proof of Theorem 1.6 to be sure that this problem is not used. Otherwise the reasoning would be circular.

7. Prove that $-(-a) = a$ for every number a.

   Answer: $a + (-a) = 0$ Theorem 1.6.

8. Prove that $-(a + b) = (-a) + (-b)$ for all numbers a and b.

A *relation* in a set is a rule that attaches to each ordered pair of elements from the set one, and only one, of the labels true or false. Equality is a relation in the set of numbers; for example, "false" goes with 1,0 as proved in Theorem 1.5; 1,0 is not in the equality relation, and this is recorded by $1 \neq 0$.

There is an order relation "less than" in our set of numbers, symbolized by $<$. As a consequence of our agreements about equality and relation, if a, b, c, and d are numbers for which $a=b$ and $c=d$, then $a<c$ implies $b<d$.

An abbreviation of a logical statement is implicitly introduced here. The symbol $a<b$ stands for a statement which may be true or false. However, it is customarily understood that the unqualified symbol $a<b$ means that $a<b$ is true.

Axioms for order follow.

O1 If a and b are numbers, then one, and only one, of $a=b$, $a<b$, $b<a$ is true.

O2 If a, b, and c are numbers for which $a<b$ and $b<c$, then $a<c$.

O3 If a, b, and c are numbers for which $a<b$, then $a+c<b+c$.

O4 If a, b, and c are numbers for which $a<b$ and $0<c$, then $ac<bc$.

To facilitate the application of these order axioms we augment the notation somewhat as follows.

$b>a$	means $a<b$.
$a \leqslant b$	means either $a<b$ or $a=b$.
$a<b<c$	means $a<b$ and $b<c$ (read b is "between" a and c).
$a \not< b$	means $a<b$ is false.

We now prove a short sequence of theorems culminating in the order relation of 0 and 1.

THEOREM 1.6 Given a number a, let $-a$ and $-b$ each stand for a negative of a. Then $-a=-b$.

Proof:

$a+(-a)=0,$	
$a+(-b)=0$	hypothesis
$(-b)+(a+(-a))=(-b)+0$	definition of operation
$((-b)+a)+(-a)=-b$	A2, A3
$0+(-a)=-b$	A1, hypothesis
$-a=-b.$	A1, A3

Therefore, once a number is chosen, there is only one number which is its negative. Axiom A4 assures us that this number has a negative. ■

THEOREM 1.7 If a and b are numbers, then $a(-c) = -(ac)$.

Proof:

$a(-c) + ac = a((-c) + c)$	D
$(-c) + c = 0$	A1, A4
$a((-c) + c) = 0$	Theorem 1.4
$a(-c) + ac = 0$	first and third steps
$a(-c) = -(ac).$	A4, Theorem 1.6[5] ■

THEOREM 1.8 If c is a number and $c < 0$, then $-c > 0$.

Proof:

$c + (-c) < 0 + (-c)$	hypothesis, O3
$0 < -c$	A1, A3, A4 ■

If a and b are numbers, then by Theorem 1.6 there exists uniquely a number $-b$, and $a + (-b)$ is a certain number by the definition of an operation. Usually, we denote this latter number by $a - b$. The three alternative symbols

$$a - b, \quad a + (-b), \quad -b + a \quad \text{(meaning } (-b) + a\text{)}$$

are equal.

THEOREM 1.9 If a, b, and c are numbers for which $a < b$ and $c < 0$, then $ac > bc$.

Proof:

$-c > 0$	hypothesis, Theorem 1.8
$a(-c) < b(-c)$	hypothesis, O4
$-ac < -bc$	Theorem 1.7
$ac - ac < ac - bc$	O3
$0 < ac - bc$	A4
$bc + 0 < bc + (ac - bc)$	O3
$bc < ac$	A1, A2, A3 ■

THEOREM 1.10 $0 < 1$.

Proof: Axiom O1 gives us a choice of just one of three possibilities. Of these, Theorem 1.5 rules out $0 = 1$. Suppose that $1 < 0$; then apply Theorem 1.9 with $a = 1$, $b = 0$, and $c = 1$ so that the hypotheses of Theorem 1.9 are fulfilled. Its conclusion for this case is

$$1 \cdot 1 > 0 \cdot 1,$$

that is, $1 > 0$, in view of Axiom M3 and Theorem 1.4. But this contradicts Axiom O1; we cannot assert both $1 < 0$ and $0 < 1$. Hence, $1 \nless 0$. The remaining possibility, $0 < 1$, is true. ■

[5] $-(ac)$ will be abbreviated to $-ac$.

1.1 PROBLEMS *(continued)*

9. Prove that if $1/a$ and $1/b$ each stand for a reciprocal of a, then $1/a = 1/b$.
   Hint: Follow the pattern of the proof of Theorem 1.6.
10. Define $a/b = a(1/b)$, and prove that $1/(a/b) = b/a$. Is there a condition necessary on what numbers a and b can stand for? (See Axiom M4.)
11. Prove that $1/(1/a) = a$ for every nonzero number a.
12. Prove that $(-a)(-b) = ab$.
    Hint: Fill in the reasons for the steps indicated by: Let $c = -b$. Then

$$(-a)(-b) = (-a)c = -ac = -a(-b) = -(-ab) = ab.$$

13. Prove that if $a \neq 0$, then either $a > 0$ or $-a > 0$, and not both.
    Hint: If $a \not> 0$, then $a < 0$. Apply Theorem 1.9 with $c = -1$.
14. Prove that $a < b$ if, and only if, $-b < -a$.

1.2 PROOFS OF THEOREMS

The proof of a theorem consists of a sequence of logical deductions, the premises for each deduction being drawn from the hypotheses of the proposition to be proved, from conclusions of steps made earlier in the proof, and from axioms and the theorems previously proved from them. The conclusion of the last deduction is the desired conclusion of the theorem.

One way to display the details of a proof is to list the steps of the proof in a column at the left of the page and to indicate in a second column, on the right, the axioms or conclusions (from whichever source they are drawn) which justify each step. This is the method that we used in the proofs of Theorems 1.1, 1.3, and 1.4, for example. As we remarked earlier, all the details are usually not necessary in order to inform a reader of the complete logical detail. Indeed, it not only saves time and type but often adds interest and clarity to point out the line of argument and to show the hard steps, leaving the more routine steps for the reader to supply.

As an example, let us rewrite the statement and the proof of Theorem 1.7.

THEOREM 1.7 $a(-c) = -ac$.

Proof: By Axiom D and Theorem 1.4, $a(-c) + ac = a(c - c) = 0$; hence, $a(-c) = -ac$ from Theorem 1.6. ■

In addition to the straightforward kind of proof we have been illustrating here, another commonly used kind is called the *indirect* proof. Theorem 1.5 is proved this way. In this method the negation of the desired conclusion is assumed and from this is deduced the negation of one of the hypotheses of the theorem, the negation of an axiom, the negation of a known true theorem, or some proposition and its own negation together. When this is done, we agree that since the negation of the conclusion has led to the contravention of a logical ground rule (that we never assert both a proposition and its negation), the negation of the negation, that is, the conclusion itself is accepted as true.

Consider Theorem 1.5 again. Our original statement of this theorem is an abbreviation for the following.

THEOREM 1.5 If the number set contains at least two elements, then $1 \neq 0$.

Proof: (method) Assume the negation of $1 \neq 0$, and conclude that the number set contains less than two elements, the negation of the hypothesis. ■

This is the simplest kind of indirect proof. If you have studied propositional calculus, a subject in logic, you recognize this as an example of the equivalence of the sentence.

p implies q

(where p and q are propositions) with the sentence

negation of q implies negation of p.

Most indirect proofs are more intricate than this, however, and we will not in general rely on formulas from formal logic.

As one more example, recall Theorem 1.10. There we deduced that the truth of $1 < 0$ necessarily carries with it the truth of $1 > 0$. But the simultaneous truth of these two propositions negates Axiom O1. Therefore, $1 < 0$ is false; that is, $0 \leq 1$ is true. The case $0 = 1$ was handled separately.

Theorem 1.10 also can be recast in the simpler form "If $<$ satisfies Axiom O1, then $0 < 1$." Then the indirect proof argues the equivalent proposition:

$0 \nless 1$ implies $<$ does not satisfy Axiom O1.

In other words,

negation of $0 < 1$ implies negation of Axiom O1.

It is one thing to read and understand a proof someone else has written, but more difficult to construct an original proof. In reading a proof you need only check each step and see that it is correct. When you build your own proof you must discover the steps that lead from hypothesis to conclusion. Unfortunately, no rules lead to this discovery. Usually we try again and again, avoiding blind alleys by free use of imagination and by experience gained in reading proofs of similar theorems.

Some proofs in this book are hard to understand at first. Usually this difficulty diminishes as soon as you get over your amazement that anyone could have invented them in the first place. This should increase your respect for the subject and not be a cause for alarm.

Constructing a complicated proof requires much more experience with mathematics than you are expected to have, and it is not the object of this book to make you an expert theorem-prover. Rather, we wish to show the logical structure of calculus and how its theorems are consequences of a few axioms, and to indicate ways by which it can be applied to important scientific problems.

1.2 PROBLEMS

1. Prove that if x and y are numbers with $xy = 0$, then either $x = 0$ or $y = 0$.
   Hint: If neither x nor y is 0, they each have a reciprocal. Multiply xy by the product of these reciprocals and deduce the negation of Theorem 1.5.
2. Find all members of each "solution" set and prove that you have them all; (use only the numbers provided by an axiom).[6]
   (a) $\{x|4x - 3 = 0\}$
   (b) $\{x|4x^2 - 3x - 1 = 0\}$
   (c) $\{x|4x^2 - 3x + 1 = 0\}$
3. Prove that if $ab > 0$, then either $a > 0$ and $b > 0$, or $a < 0$ and $b < 0$.
   Hint: If $a > 0$ and $b \not> 0$, then either $ab = 0$ by Theorem 1.4 or $ab < 0$ by Theorems 1.4 and 1.9. This negates the hypothesis. The other cases are handled in a similar way.
4. Prove that $1/a > 0$ if, and only if, $a > 0$.
5. Prove that $1/1$, the reciprocal of 1, is 1.
6. Prove that there is no reciprocal of 0.
   Hint: If it did exist, then $(1/0) \cdot 0 = 1$, and with Theorem 1.4 this negates Theorem 1.5.
7. Prove that the symbol $0/0$ is available to stand for any number.
   Hint: The definition $0/0 = 0(1/0)$ of Problem 10, Section 1.1, would require the undefined $1/0$; therefore there is no restriction on the use of this symbol $0/0$. Henceforth this symbol will not stand for a number.

[6] See Appendix I for an explanation of set notation.

1.3 DEFINITIONS

From the point of view of logic, definitions are not necessary. One usually defines a circle, say, to be the locus of points equidistant from a given point. In a geometry problem involving a circle one could write "locus of . . ." whenever he needed to, and never use the word "circle." It might amuse you to try this after we present several definitions. Delete a few definitions, and rewrite theorems and their proofs without them. You can do it with no loss of rigor in the reasoning, but you will discover a great loss of precision in your thinking and an unbearable boredom with the verbiage. Definitions are a practical necessity.

The word "robin" is in the vocabulary of the ornithologist, although he could write down the defining characteristics of the creature whenever he referred to it and avoid the use of "robin." In mathematics, too, we expand the mathematical vocabulary by adding a new term when it serves to condense a lengthy description of some mathematical object. This often confuses the layman because a mathematician usually picks a common word and uses it for his own purposes. As we shall see, a *limit* is a number in a specific context; this word does not have this meaning in ordinary discourse.

As a case in point, choose a number a. Axiom A4 asserts that there is a number $-a$, that we know there is only one such number, and that $a + (-a) = 0$. Let us construct the following definition.

DEFINITION 1.1 A number a is *positive* in case $a > 0$ and is *negative* in case $a < 0$. If $a \not> 0$, then a is *nonpositive;* that is, $a \leq 0$. If $a \not< 0$, then a is *nonnegative;* that is, $a \geq 0$.

In Problem 13, Section 1.1, you proved that for any number a, exactly one of the statements $a \geq 0$ or $-a > 0$ is true. This justifies the next definition.

DEFINITION 1.2 If a is a number, then the *absolute value* of a is $|a|$, where

$$\begin{aligned} |a| &= a, \quad \text{when } a \geq 0 \\ |a| &= -a, \quad \text{when } a < 0. \end{aligned}$$

Thus, $|a|$ is a symbol that stands for a or $-a$, whichever is nonnegative. For example, if $a = 3$ (nonnegative, itself), then $|a| = |3| = 3$. If $a = -3$ (so that $-(-3)$ is nonnegative), then $|a| = |-3| = -(-3) = 3$.

Note here that every nonzero number has a negative of itself that goes with it. On the other hand, some numbers, those which are less than 0, are designated specifically as negative numbers.[7]

One very useful algebraic formula is contained in the next theorem. It is called the *triangle inequality* because of its geometric interpretation in the complex number system.

THEOREM 1.11 If a and b are numbers, then

$$|a+b| \leqslant |a| + |b|.$$

Proof: If $a+b \geqslant 0$, then by the definition of absolute value and by Problem 2, Section 1.3, $|a+b| = a+b \leqslant |a| + |b|$. If $a+b<0$, it follows similarly with Problem 4, Section 1.3, that $|a+b| = -(a+b) = -a-b \leqslant |-a| + |-b| = |a| + |b|$. ■

1.3 PROBLEMS

1. Compute $|a|$ for each of $a = 1, -1, 2, -2, 5$, and -5. Verify for each case that $a \leqslant |a|$ and that $-|a| \leqslant a$. Test each of the formulas proved in the next problems with several numbers to see how they work.
2. Prove that for any number a, $-|a| \leqslant a \leqslant |a|$.
   Hint: If $a \geqslant 0$, then $-|a| = -a \leqslant a = |a|$, and if $a < 0$, then $-|a| = -(-a) = a < -a = |a|$.
3. Prove that if $a \neq 0$, then $|a| > 0$.
   Hint: By Axiom O1, either $a > 0$ so that $|a| = a > 0$, or $a < 0$ so that $|a| = -a > 0$.
4. Prove that $|a| = |-a|$ for any number a.
5. Prove that for any numbers a and b, $|a| \cdot |b| = |ab|$.
   Hint: If $ab > 0$, then either $a > 0$ and $b > 0$ so that $|a| \cdot |b| = a \cdot b = |ab|$, or $a < 0$ and $b < 0$ so that $|a| \cdot |b| = (-a)(-b) = ab = |ab|$. Treat the remaining cases, $ab = 0$ and $ab < 0$, similarly. There are no other cases.
6. Prove that for any numbers a and b

   $$|a-b| \leqslant |a| + |b|.$$

   Hint: By the triangle inequality,

   $$|a-b| = |a+(-b)| \leqslant |a| + |-b| = |a| + |b|.$$

7. Prove that for any numbers a and b, $|a| - |b| \leqslant |a-b|$.

[7] Some textbooks emphasize the distinction between a negative number and the negative of a number by using a raised hyphen for the former, ⁻3, and an ordinary hyphen for the latter, -(⁻3), calling this the *opposite* of ⁻3.

Hint: By the triangle inequality,

$$|a| = |(a-b) + b| \leqslant |a-b| + |b|.$$

8. Prove a strengthened form of the inequality of Problem 7, namely,

$$||a| - |b|| \leqslant |a-b|.$$

Hint: Either $|a| \geqslant |b|$, in which case

$$||a| - |b|| = |a| - |b| \leqslant |a-b|$$

by Problem 7, or $|a| < |b|$, in which case

$$||a| - |b|| = |b| - |a| \leqslant |b-a| = |a-b|$$

by Problems 4 and 7.

9. Solve the equation $|x+1| = 3$.
Hint: From the definition of the absolute value of a number, either $|x+1| = x+1$ or $|x+1| = -(x+1)$. Thus the only two possibilities are $x+1=3$ and $-(x+1)=3$. The first gives $x=2$ and the second, $x=-4$, which are solutions as can be seen by substitution in the original equation.

10. Solve the following equations.
(a) $|2-x| = 1$
Answer: $x=1$ and $x=3$
(b) $|2-x| = 4$
(c) $|4x+1| = 7$
(d) $|x+1| = x$
Answer: no solution
(e) $|x-1| = 2x+4$
(f) $|x+1| = |x-2|$
Answer: $x = \frac{1}{2}$
(g) $|2x+3| = |4-x|$
(h) $|3x+2| + |2x+3| = 0$

11. Solve the following equations.
(a) $|x+1|^2 - 3|x+1| + 2 = 0$
Hint: This is a quadratic equation in $|x+1|$. Solving, we have $|x+1| = 1$ or $|x+1| = 2$. The first gives $x=0$ or -2 and the second, $x=1$ or -3.
(b) $|x-1|^2 - 5|x-1| + 6 = 0$
(c) $|2x+1|^2 + |2x+1| - 2 = 0$
Answer: $x=0$ and $x=-1$
(d) $|1-x|^2 - 2|1-x| = 0$
(e) $|x+3|^2 + 4|x+3| + 3 = 0$
(f) $|x+2|^3 = 8$

12. Solve the following pairs of simultaneous equations.
   (a) $|x| = 1$ and $|x - 2| = 1$
   (b) $|x - 3| = 2$ and $|x - 7| = 2$
   (c) $|x - 3| = 2$ and $|x + 3| = 1$
   (d) $|4 - x|^2 + |4 - x| - 6 = 0$ and $|3 - x| - 3 = 0$

1.4 INTERVALS

DEFINITION 1.3 If a and b are numbers, with $a < b$, the *open interval* (a,b) is the set of all numbers between a and b.

If we wish to designate a set of numbers consisting of an open interval (a,b) with either a or b adjoined to the set, we use the symbols $[a,b)$ or $(a,b]$, respectively. The bracket is next to the number added on. Thus, for example,

$$(a,b] = \{x | a < x \leqslant b\}$$

in set notation. The set

$$[a,b] = \{x | a \leqslant x \leqslant b\}$$

is called the *closed interval* $[a,b]$. Occasionally we use such symbols as

$$(a,\infty) = \{x | a < x\}$$
$$(-\infty,\infty) = \{x | x \text{ is a number}\}$$

and others like them.

We assume that you are familiar with the *number line*, a certain one-to-one correspondence between points of a line and the number set we are describing in this chapter. The precise relationship of the geometric point set, the line, to the number system is difficult to prescribe in detail. Since this is not a book about geometry, we will simply conceive of the number line as a long ruler with an ordinary scale to picture and keep track of the number relations which are our principal concern. The practical connection between the line and the number system is that if a point O is assigned the number 0 and a point U is assigned the number 1, then a scale can be assumed so that if the number a goes with the point A and b goes with B, then $b - a$ measures the distance from A to B. This is shown in Figure 1.1.

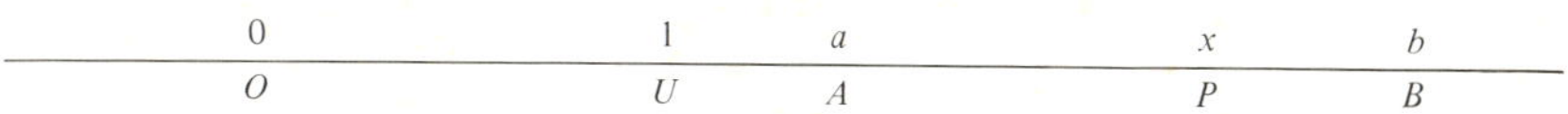

Figure 1.1

Also, if x goes with P, and $a < x < b$, then the point P appears to be between A and B on the line.

A direction, or sense, on the line is positive if it is the same as the direction from O to U, negative if it is opposite. If $b - a$ is positive, then the segment $\overline{AB}$ has the same direction as $\overline{OU}$; if $b - a$ is negative, $\overline{AB}$ has direction opposite to $\overline{OU}$. If a second segment is determined by the number c which goes with point C, and d which goes with D, then $|b - a| = |d - c|$ if, and only if, the segment $\overline{AB}$ is congruent to the segment $\overline{CD}$.

With these conventions assumed, the open interval (a,b) has the diagrammatic representation shown in Figure 1.2.

Figure 1.2

The parentheses on the line indicate the omission of the endpoints. We represent the interval $(a,b]$ by the diagram in Figure 1.3.

Figure 1.3

When a number x goes with a point P, we call x the *coordinate* of P, and record this correspondence by $P(x)$.

1.4 PROBLEMS

1. Draw a line and place points $O(0)$ and $U(1)$ on it. We call this accomplishment placing a coordinate system on the line. Locate points whose coordinates are $1, 2, -3, 4, -8, \frac{1}{2}, \frac{1}{3}$, and $-\frac{1}{4}$.
2. Assume that $U(1)$ is to the right of $O(0)$ on a horizontal line. Find the coordinate of each point P described below.
   (a) P is 2 units to the right of $Q(3)$.
   (b) P is 2 units to the left of $Q(6)$.
   (c) P is 1 unit to the left of $Q(-4)$.
   (d) P is twice as far from $O(0)$ as is $Q(3)$, and is on the opposite side of O.
   (e) $Q(7)$ is 3 units to the right of P.
3. Find the distance between each pair of points listed below.
   (a) $O(0)$, $P(4)$
   (b) $O(0)$, $P(-4)$
   (c) $P(4)$, $Q(-4)$
   (d) $P(3)$, $Q(-1)$
   (e) $P(-2)$, $Q(-3)$

4. A definition of distance from P to Q is suggested by the language used in the preceding two problems.

DEFINITION Let a coordinate system on a line be given. The directed *distance from* $P(x)$ to $Q(y)$ is the number $y - x$. The *distance between* $P(x)$ and $Q(y)$ is the number $|y - x|$. If $P(2)$ and $Q(y)$, find y if the directed distance from P to Q is

(a) 1 (b) 3 (c) $\frac{1}{2}$ (d) -1 (e) -3.

5. If $P(-3)$ and $Q(y)$, find y if the distance between P and Q is
(a) 3 (b) $\frac{1}{2}$ (c) 2 (d) 5.
6. Prove that if P, Q, and R are points on a line with a coordinate system, the sum of the distances from P to Q, from Q to R, and from R to P is 0. The distance *from one point to another* always refers to the directed distance.
7. Solve the inequality $-1 < 3 - 2x < 0$.
Hint: $-3 - 1 < 2x < -3$
$\frac{3}{2} < x < 2.$
Therefore, the solution set is the open interval $(\frac{3}{2}, 2)$.
8. Solve each of the following inequalities and express the solution set as an interval or combination of intervals.
(a) $2 \leqslant x - 2 < 4$
Answer: $4 \leqslant x < 6$, or $[4,6)$
(b) $8 \leqslant 4x - 2 \leqslant 9$
(c) $-3 \leqslant 3x + 3 \leqslant 0$
(d) $2 \leqslant 1 - x \leqslant 5$
Answer: $-4 \leqslant x \leqslant -1$
(e) $1 \geqslant 2 - 3x > -4$
(f) $\dfrac{1}{2} \leqslant \dfrac{10 - x}{3} \leqslant 2$
(g) $2x + 3 \geqslant 1$
Answer: $[-1,\infty)$
(h) $3 - 2x \leqslant 4$
(i) $\dfrac{1}{x+1} < 2$
Answer: $-\frac{1}{2} < x$ and $x < -1$
(j) $\dfrac{1}{2x+3} < 4$
(k) $\dfrac{x+1}{2x+3} < 4$
(l) $1 < \dfrac{1}{x+1} < 3$
Answer: $-\frac{2}{3} < x < 0$

(m) $0 \leqslant \dfrac{1}{1-x} < 1$

(n) $\dfrac{x+1}{4x-3} < 1$

Answer: $(-\infty,\frac{3}{4})$ and $(\frac{4}{3},\infty)$

(o) $\dfrac{x-1}{2x+1} < 2$

(p) $\dfrac{x-1}{2x+1} < 0$

Answer: $-\frac{1}{2} < x < 1$

(q) $\dfrac{2-3x}{3+2x} < 0$

(r) $(2-3x)(3+2x) < 0$

(s) $(x-2)(3+x) > 0$

9. Solve the inequality $|x-1| < 4$.

Answer: If $x-1 \geqslant 0$, then $x \geqslant 1$ and $|x-1| = x-1$. Therefore, $x-1 < 4$ and $x < 5$, so $1 \leqslant x < 5$. If $x-1 < 0$, then $x < 1$, $x-1 > -4$, and $x > -3$, so $-3 < x < 1$. Suppose that a number x belongs to the union of intervals $(-3,1) \cup [1,5)$. If x is in $(-3,1)$ so that both $x < 1$ and $x > -3$, that is, $x-1 < 0$ and $x-1 > -4$ hold, x will satisfy $|x-1| < 4$. If not, then x is in $[1,5)$ and, again, x will satisfy $|x-1| < 4$. Therefore, $|x-1| < 4$ if, and only if, x is in $(-3,5)$.

10. Solve each of the following inequalities.

(a) $|x-3| < 2$

(b) $|5+x| < 7$

(c) $|4x-8| < 12$

11. Prove that if $\delta > 0$ and $|x-a| < \delta$, then the solution set of the inequality is $(a-\delta, a+\delta)$.

12. Use Problem 11 to write the solution set of each of the following.

(a) $|x-4| < 6$

(b) $|x+4| < 6$

Hint: $4 = -(-4)$.

(c) $|3-x| < 2$

Hint: $|3-x| = |x-3|$.

(d) $|5x-20| < 10$

Hint: $|5x-20| = 5|x-4|$.

13. Prove that if $\delta > 0$ and $|x-a| > \delta$, then the solution set of the inequality is the union of intervals $(-\infty, a-\delta) \cup (a+\delta, \infty)$.

14. Use Problem 13 to solve each inequality of Problem 12, with $<$ replaced by $>$.

15. Prove that if $e > 0$ and $f > 0$, the solution set of the pair of inequalities $|x-a| < e$ and $|x-b| < f$ is the intersection of intervals $(a-e, a+e) \cap (b-f, b+f)$.

16. Prove that if $\delta < 0$, the solution set of $|x-a| < \delta$ is empty.

1.5 THE LEAST UPPER BOUND AXIOM

Axiom M3 asserts the existence of a number "one," and Axiom M1, used in the proof of Theorem 1.2, implies that this number is unique. With this number we describe a way to generate the whole numbers, by adding ones, starting with Problem 16 of the review problems. In a somewhat similar way Axiom M4, by asserting the existence of reciprocals, can be used to generate the rational numbers. We presume that whole numbers and rational numbers, how they are related to each other, and how to compute with them are understood.

When we solve an algebraic equation, we typically work with one or more letters (*place holders* or *unknowns*) which we assume stand for numbers, and we operate with these letters under rules dictated by axioms. This is the practical algebra we do. All of the familiar methods for drawing conclusions in algebra can be shown to be logical consequences of the axioms; indeed, if a method is not such a logical consequence, it is incorrect.

In some of the problems you practiced verifying that certain algebraic methods which you already knew are, in fact, implied by the axioms. We will not ask you to complete the project, since it would exhaust both your patience and ours. But you should hold yourself accountable to the axioms through theorems derived from them for any algebraic fact or technique you use.

We now ask if there exist numbers other than rational numbers, that is, irrational numbers. Apparently, we have as yet no axiom that asserts the existence of an irrational number or that hints at how one might arise. We do know, by Problem 1, Section 1.5, that the equation $x^2-2=0$ has no rational solution, so we might put down an axiom which asserts that it does have some solution. With such an axiom we would still face the question of the existence of irrationals other than the "quadratic" irrational introduced as a solution of this equation. In what follows we use the equation $x^2-2=0$ to lead to the *least upper bound* axiom, and this axiom generates all the irrational numbers in the real number system.

At this point we must be quite candid about the way in which we mean to use the equation $x^2-2=0$. The rational number system does have the properties specified by the axioms we have listed previously. However, this assertion cannot be convincing unless rational numbers are precisely constructed so that you know exactly what they are. We will not take the time now to do this. In fact, the rational number system possesses one property which is not implied by the axioms. This is the *Archimedean* property, which says

> If r and s are rational, with $0<r<s$, then there is a rational integer n such that $s<nr$.

There are number systems in which the axioms are true but which are non-Archimedean. Therefore, in the following illustration we draw upon your experience, which for the moment is outside the theory we have built, just to indicate why one more axiom is useful.

Specifically, we will assume that each rational number has a decimal representation and that for any particular rational number there is a power of ten, 10^n, which exceeds that number. We will rely on your acquaintance with decimals. Fortunately, once the real number system is specified, there can be found within it an Archimedean rational number system with the properties sufficient to keep our illustration based on $x^2 - 2 = 0$ honest; for example, see Problem 13 of the review problems.

A decimal computer can use rational numbers only, and we may visualize a program for a computer as we search for a solution of $x^2 - 2 = 0$. With the axioms at hand, the computer's numbers are ours. Notice that 1 is too small to be a solution, 2 is too large, and that 1.4 is too small, 1.5 is too large. Calling these steps the zero[th] and first trials, respectively, we can generalize the process. Let $\underline{m}_n$ and $\bar{m}_n$ be the lower and upper numbers found at the nth trial, and construct the sets.

$$\underline{M} = \{1, 1.4, 1.41, \ldots, \underline{m}_n, \ldots\}$$
$$\bar{M} = \{2, 1.5, 1.42, \ldots, \bar{m}_n, \ldots\}$$

in which $\underline{m}_n^2 < 2$, $\bar{m}_n^2 > 2$, $\bar{m}_n - \underline{m}_n = 1/10^n$, $n = 0, 1, 2, \ldots$, and no number in either set has a square equal to 2, by Problem 1, Section 1.5. Furthermore, $\underline{m}_n \leqslant \underline{m}_{n+p}$ and $\bar{m}_n \geqslant \bar{m}_{n+p}$ for each n and all p, and each number in $\underline{M}$ is less than any number in $\bar{M}$. See Figure 1.4.

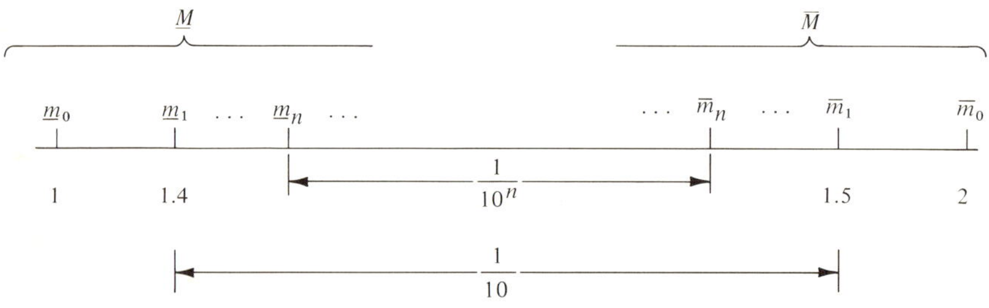

Figure 1.4

DEFINITION 1.4 A number u is an *upper bound* of a set M of numbers if $m \leqslant u$ for each number m in the set M.

In our example each number in $\bar{M}$ is an upper bound of the set $\underline{M}$. Clearly, a set may have many upper bounds (as $\underline{M}$ does) or no upper bound (such as the set of all whole numbers). An upper bound of a set may belong to it; for example, 3 is an upper bound of $\{1,2,3\}$—or it may not, as 1 is an upper bound of $\{\frac{1}{2},\frac{3}{4},\frac{7}{8}, \ldots\}$.

DEFINITION 1.5 A number l is a *least upper bound*[8] of a set M of numbers if both (a) and (b) are true, where

(a) l is an upper bound of M,
(b) if u is an upper bound of M, then $l \leqslant u$.

In our example we have seen no lub $\underline{M}$, but we soon return to search for it. In Problem 7, Section 1.5, you will prove that if a set of numbers has a least upper bound, it has only one. A least upper bound of a set may belong to it, as 3 belongs to $\{1,2,3\}$, or it may not, as $1 = \text{lub}\{\frac{1}{2},\frac{3}{4},\frac{7}{8}, \ldots\}$. If a set has no upper bound, it has no least upper bound (see (a) of Definition 1.5). Interestingly, though, a set of numbers in certain number systems may have an upper bound but no least upper bound. This is the case for the set $\underline{M}$ in the rational number system, as we now demonstrate.

Consider the following two propositions, which are proved in the problems below, to be true.

PROPOSITION 1.12 If the set $\underline{M}$ has a least upper bound $l = \text{lub } \underline{M}$, then $l^2 = 2$.

Notice carefully that this proposition does not assert that lub $\underline{M}$ exists; it only asserts that *if* it did, *then* the square of that lub would have to be 2.

PROPOSITION 1.13 If, for some $x > 0$, $x^2 = 2$, then $x = \text{lub } \underline{M}$.

Again, we do not say $x^2 = 2$ has a positive solution; we only assert that *if* it did, *then* that solution would have to be lub $\underline{M}$.

We conclude, after proving these propositions, that if $\underline{M}$ has a least upper bound, it cannot be a rational number; our decimal computer will never find it. For if it were rational, there would be a rational number whose square is 2 (Proposition 1.12), and this is false. Also, this hypothetical $l = \text{lub } M$ is the only possible positive solution of $x^2 - 2 = 0$ (Proposition 1.13). So without the existence of l, a solution of the equation is hopeless. This impels us to assert another axiom to strengthen the problem-solving power of the number system.

LEAST UPPER BOUND AXIOM If a nonempty set M has an upper bound, then M has a least upper bound.

This is an existence axiom, and like other existence axioms we have

[8] The least upper bound of M is written lub M.

remarked, it serves to generate numbers. In our example it generates lub $\underline{M}$, which we have seen has 2 for its square. Although this number cannot be written as a decimal, it exists in a number system for which this axiom is assumed. The usual symbol for it is $\sqrt{2}$. Usually (not always) when the existence of a number rests on this axiom, there is no standard symbol for it (no decimal whenever it is irrational), and we denote it by a convenient letter.

DEFINITION 1.6 The *real number system* is a number system which satisfies Axioms A1, A2, A3, A4, M1, M2, M3, M4, D, O1, O2, O3, O4, and the least upper bound axiom.

Only rarely do we appeal to the least upper bound axiom. But when we do, it is at a place central in, and basic to, the theory of calculus.

1.5 PROBLEMS

1. Use properties of whole numbers to prove that there is no rational $\sqrt{2}$ (a number a is rational if there exist whole numbers p and q so that $a = p/q$).
   Answer: $x = \sqrt{2}$ is a positive solution of $x^2 = 2$. If $x = u/v$, then $x = p/q$, where p and q are not both even. Therefore, $(p/q)^2 = 2$ and $p^2 = 2q^2$. If p were an odd number, say $p = 2n + 1$, then $p^2 = 2(2n^2 + 2n) + 1$, which is odd also. Hence, since p^2 is even, $p = 2m$ and $2q^2 = 4m^2$ and $q^2 = 2m^2$. As before, q cannot be odd, so q is even, contradicting the choice of p and q.
2. Prove that $\sqrt{3}$, $\sqrt[3]{3}$, and $\sqrt{10}$ are not rational numbers.
3. Prove that if $l = \text{lub } \underline{M}$, then $l^2 = 2$.
   Proof: Suppose that $l^2 < 2$; then $2 - l^2 > 0$. Take n so that

$$\frac{4}{10^n} < 2 - l^2.$$

Then since both $\underline{m}_n$ and $\bar{m}_n$ are less than 2,

$$\bar{m}_n{}^2 - \underline{m}_n{}^2 = (\bar{m}_n - \underline{m}_n)(\bar{m}_n + \underline{m}_n) < \frac{4}{10^n}.$$

Therefore,

$$\underline{m}_n{}^2 > \bar{m}_n{}^2 - \frac{4}{10^n} > 2 - (2 - l^2) = l^2$$

and

$$(\underline{m}_n - l)(\underline{m}_n + l) = \underline{m}_n{}^2 - l^2 > 0.$$

This means that $\underline{m}_n > l$, so l is not an upper bound of $\underline{M}$.

Suppose that $l^2 > 2$; then $l^2 - 2 > 0$. Take n so that

$$\frac{4}{10^n} < l^2 - 2.$$

Then, as before,

$$\bar{m}_n^2 - \underline{m}_n^2 < \frac{4}{10^n}$$

and

$$\bar{m}_n^2 < \underline{m}_n^2 + \frac{4}{10^n} < 2 + (l^2 - 2) = l^2.$$

Therefore,

$$(\bar{m}_n - l)(\bar{m}_n + l) = \bar{m}_n^2 - l^2 < 0,$$
$$\bar{m}_n < l.$$

However, for all p,

$$\underline{m}_p \leqslant \bar{m}_n < l,$$

which means that l is not a least upper bound of $\underline{M}$. ■

4. Prove that if $x^2 = 2$ and $x > 0$, then $x = \text{lub } \underline{M}$.

Proof: For any $\underline{m}_n$ in $\underline{M}$, $\underline{m}_n^2 < 2 = x^2$; that is, $\underline{m}_n < x$, and x is an upper bound of $\underline{M}$.

Let x_1 be an upper bound of $\underline{M}$. If $x_1 < x$, then $x_1^2 < x^2 = 2$, and take n so that

$$\frac{4}{10^n} < 2 - x_1^2.$$

Now $\underline{m}_n^2 \leqslant x_1^2$, and as in Problem 3,

$$\bar{m}_n^2 - \underline{m}_n^2 < \frac{4}{10^n}$$

so that

$$\bar{m}_n^2 < \underline{m}_n^2 + \frac{4}{10^n} < x_1^2 + (2 - x_1^2) = 2,$$

contradicting the choice of $\bar{m}_n$. Therefore, $x \leqslant x_1$. ■

5. Prove that $\underline{M}$ contains no greatest number.
6. Prove that if a set of numbers has an upper bound, it has infinitely many of them.
7. Prove that a set of numbers can have at most one least upper bound.

Hint: Let $l = \text{lub } M$ and $m = \text{lub } M$. Then l is an upper bound of M and $m \leqslant l$. Also, m is an upper bound of M and $l \leqslant m$. Therefore, $l = m$.

8. Let M be a set of numbers and $l = \text{lub } M$. Let N be the set of numbers obtained by multiplying each number in M by the positive number t. Prove that $tl = \text{lub } N$.
9. Let $l = \text{lub } M$. Prove that if $\epsilon > 0$, there is a number m in M for which $l - \epsilon < m \leqslant l$.

   Hint: If there were no such number m, then you could show that $l - (\frac{1}{2})\epsilon$ would be an upper bound of M, less than l.
10. DEFINITION b is a *lower bound* of a nonempty set M if whenever m is in M, $b \leqslant m$. A number g is a *greatest lower bound* of M ($g = \text{glb } M$) in case

    (a) g is a lower bound of M,
    (b) and if b is a lower bound of M, then $b \leqslant g$.

    Give examples of sets of numbers which have

    (a) both an upper and a lower bound;
    (b) an upper bound but no lower bound;
    (c) a lower bound but no upper bound;
    (d) neither an upper nor a lower bound.

    Include cases for which the bounds belong to, and do not belong to, the sets.
11. Prove that if a set of numbers has a lower bound, it necessarily has a greatest lower bound. (It is not necessary to impose a separate greatest lower bound axiom on the real number system to get this property.)

    Answer: (if the set is not empty)

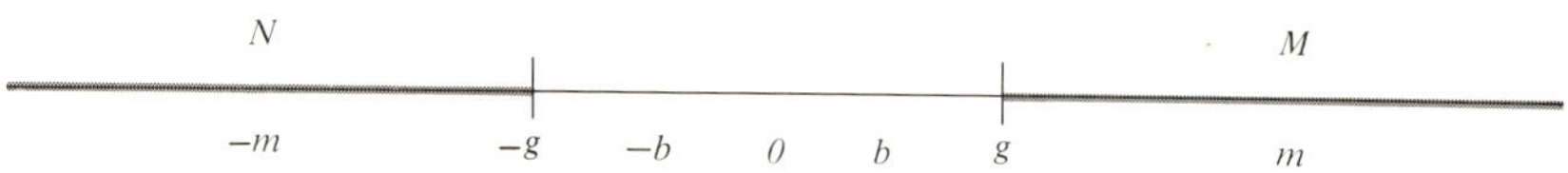

Figure 1.5

Let b be a lower bound of a set M and denote a member of M by m. Consider the set N of negatives $-m$ of the numbers in M. (Refer to Figure 1.5 to keep track.) For any $-m$ in N, m is in M with $b \leqslant m$; hence, $-m \leqslant -b$ and $-b$ is, therefore, an upper bound of N. By the least upper bound axiom, N has a least upper bound—call it $-g = \text{lub } N$. We prove that $g = \text{glb } M$ as follows. If m is in M, then $-m$ is in N, and so $-m \leqslant -g$ and $g \leqslant m$, showing that g is a lower bound of M.

Suppose that b is any lower bound of M; take any $-m$ in N, which puts m in M with $b \leqslant m$; that is, $-m \leqslant -b$, and $-b$ is an upper bound of N. But $-g \leqslant -b$, and since this implies $b \leqslant g$, we have $g = \text{glb } M$.

REVIEW PROBLEMS

1. Prove that if $ab > 0$ and $a < b$, then $1/a > 1/b$.
2. Prove that if $a > 0$ and $b > 0$, then $a + b \geqslant 2\sqrt{ab}$.
3. If $a < b$ and $c < d$, what can you say about $a - c$ and $b - d$? What can you say about $a - d$ and $b - c$?
4. Prove that $|a/b| = |a|/|b|$.
5. When is $|a - b| \leqslant |a| - |b|$?
6. Prove that for any number a, $-|a| \leqslant a \leqslant |a|$.
7. When is $|a + b| < |a| + |b|$?
8. Prove that $\frac{1}{2}(|a| + a) = a$ if $a \geqslant 0$. What is the value when $a < 0$?
9. Prove that if a and b are any two rational numbers, then there is a rational number between them.
10. Show that the following numbers are irrational:

$$\sqrt[3]{5}, \qquad \sqrt[12]{2}, \qquad \sqrt[3]{2}.$$

11. Is $\sqrt{2} + \sqrt{3}$ rational or irrational?
Hint: If $\sqrt{2} + \sqrt{3}$ is rational, so is its square. Then $\sqrt{6}$ is rational.
12. Show that if $2^a = 3$, then a is irrational.
Hint: Use what you know about the definition of $2^{p/q}$, p,q integers.
13. Show that if $a > 0$ and $b > 0$, then there is an integer n such that $na > b$.
Hint: Use an indirect proof and consider the lub of all numbers of the form na, where n is an integer.
14. Show that if $a > 0$, there is an integer n such that $1/n < a$.
15. Let M and N be two collections of numbers. Let a be the least upper bound of M, and let b be the least upper bound of N. Let P be the collection of all numbers which can be obtained by adding a number in M and a number in N. Show that $a + b$ is the least upper bound of P.
16. Prove that $1 + 1$ is neither 0 nor 1; therefore, $1 + 1$ is a new number, say 2.
17. Prove that $2 + 1$ is none of 0, 1, or 2; therefore, $2 + 1$ is a new number, say 3.
18. Define a subset M of the set R of real numbers to be *inductive* in case 1 belongs to M, and $x + 1$ belongs to M whenever x does. Let $\mathscr{S} =$ {all inductive subsets M of the set R of real numbers}. Which of the following sets is in $\mathscr{S}$?

(a) $\{1,\frac{1}{2},2,\frac{3}{2},3,\frac{5}{2}, \ldots\}$
(b) $\{-2,-1,0,1,2, \ldots\}$
(c) $\{\ldots,-2,-1,0,1\}$
(d) R
(e) {all nonnegative real numbers}
(f) $\{1,3,5,7, \ldots\}$
(g) $\{1,1\frac{1}{2},1\frac{3}{4},1\frac{7}{8}, \ldots ,2,2\frac{1}{2},2\frac{3}{4},2\frac{7}{8}, \ldots ,$
$3,3\frac{1}{2},3\frac{3}{4},3\frac{7}{8}, \ldots$, and so on$\}$

19. Define the set N of whole numbers to be the smallest inductive subset of the set R of real numbers; that is, $N = \cap \mathscr{S}$, the intersection of all sets in $\mathscr{S}$. Show that
    (a) $\mathscr{S}$ is not empty, therefore N exists as a set;
    (b) N is not empty, therefore by this definition there is a whole number;
    (c) N itself is an inductive subset of R.
20. Prove the following three properties of N.
    (a) If M is an inductive set of real numbers, then $M \supset N$.
    (b) Call x a *predecessor* of $x + 1$. The number 1 has no predecessor in N.

    Answer: Suppose that 0 belongs to N. Since N is inductive, $N - \{0\}$ is inductive also, and belongs to $\mathscr{S}$. Therefore, $N - \{0\} \supset N$ by (a), a contradiction.

    (c) No whole number has more than one predecessor.
21. Prove that 1 is the only whole number without a predecessor in N.
    Answers: Let $M = \{1$, and all whole numbers with predecessors$\}$. Then if x is in M, $x + 1$ is a whole number with predecessor x, so $x + 1$ is in M, and M is inductive. By Problem 20(a), M contains every whole number.
22. Prove that there is no whole number between 1 and $1 + 1$.
23. Prove that the sum of the first n odd whole numbers is n^2.
    Answer: Let $M = \{$all whole numbers for which the proposition is true$\}$. Since $1 = 1^2$, and if $1 + 3 + \ldots + (2n - 1) = n^2$, then $1 + 3 + \ldots + (2n - 1) + (2n + 1) = n^2 + (2n + 1) = (n + 1)^2$, M is inductive. By Problem 20(a), M contains all whole numbers.
24. Use the method of Problem 23 to establish the standard formulas for the sums of arithmetic and geometric progressions, namely,

$$a + (a + d) + (a + 2d) + \ldots + (a + (n - 1)d) = n\left(a + \frac{n-1}{2}\,d\right)$$

$$a + ar + ar^2 + \ldots + ar^{n-1} = \frac{a}{1-r}(1 - r^n).$$

Chapter 2
FUNCTIONS

2.1 INTRODUCTION

This chapter contains a definition of a function, and many examples of functions. These are the principal objects of attention for the remainder of the book. Notice that the ingredients that go into this definition and into its applications are the real number system, a little bit about sets, and some simple logic.

We do not propose to explore at all extensively the connections between mathematical objects and nonmathematical things which may behave like them. This would take us into applied mathematics and would presuppose a knowledge of the calculus we propose to construct. However, we should call attention to some things outside of mathematics which have given a direction to the calculus and from which the mathematical objects have been distilled.

2.2 DEFINITION OF A FUNCTION

DEFINITION 2.1 A *function* is the pair of things:

(a) a nonempty set of real numbers,
(b) and a rule that assigns to each number of that set one, and only one, real number.

The nonempty set referred to in this definition is called the *domain* of the function, and the set of numbers assigned by the rule is called the *range*, or *image set*, of the function.[1]

A particular function can be described in any one of many ways. We will illustrate a few such ways. Frequently we discover that a function is involved when numerical measurements are being made.

If two thermometers measure the same temperature, one with a Fahrenheit scale and the other with a Centigrade scale, then the Fahrenheit reading assigned to a given Centigrade reading is obtained by multiplying the Centigrade reading by $\frac{9}{5}$ and adding 32 (this is a rule). The domain of the corresponding function depends on how long the centigrade scale is. If this scale runs from 0 to 100, the domain of the function is the interval [0,100] and the range is the interval [32,212]. For each Centigrade scale there is defined a unique function. The rule for each such function is the same and is most concisely formulated by the familiar equation $F = \frac{9}{5}C + 32$.

If the domain of a function is a reasonably small set of numbers, the function can be described accurately by a geometrical graph. For example, in each of five consecutive years a certain number of students failed a particular course. A function is generated by this situation, and it is given by the graph in Figure 2.1.

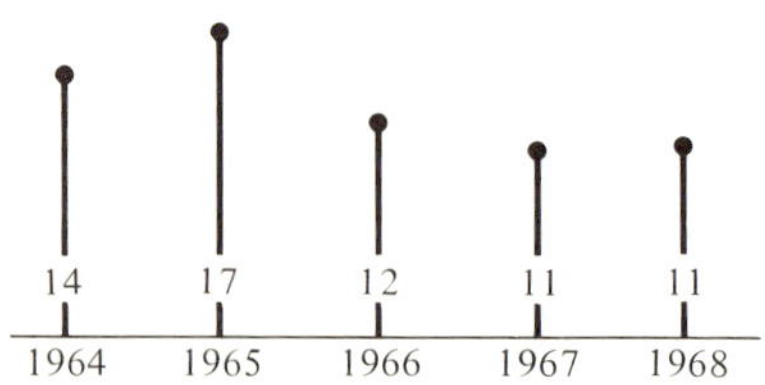

Figure 2.1

[1] More generally, functions can be defined whose domains and ranges contain other things then real numbers. The functions that now concern us are often called *real valued functions of a real variable*.

The domain and rule are clear from the picture. An equally precise description of this function is displayed by the table in Figure 2.2.

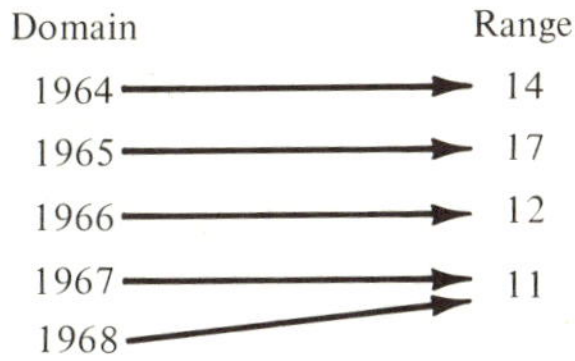

Figure 2.2

The graph suggests a trend in the number of failures more vividly than does the table. Graphs can suggest properties of functions that may be obscure in formulas and tables.

But if the domain of a function is an infinite set, neither a graph[2] nor a table can completely settle a question about a function it represents, usually. You are familiar with the sine wave as the graph of the sine function. The domain of this function is the set of real numbers, and its rule is rather complicated (involving the measurement of line segments and angles or arc lengths). The table of values of the sine function (numbers in the range) which is printed in a book of numerical tables is simply wrong for most entries, except for particular ones such as $\sin \pi/2 = 1$. These wrong entries are called *approximations*. Answers to such questions as, "Does the sine curve really cross the axis at a 45° angle?" or "Is $\sqrt[5]{0.9}$ in the range?" are inaccessible to the graph or to the table. We will find these answers soon in the calculus.

2.3 NOMENCLATURE FOR FUNCTIONS

The word *nomenclature* has to do with names for things. We need to be able to name functions and certain other objects related to them, such as a domain and a range. Suppose that we use lowercase letters $f, g, h, \phi, \ldots,$ as names of functions. Then the symbols dom f, dom $g, \ldots,$ will stand for their respective domains.

If f is a function and x is a number in dom f, there is a number in the range of f assigned to x. We denote this range number by the symbol $f(x)$, and we speak of $f(x)$ as the image of x in f or the value of f at x. Thus

$$\text{range of } f = \{f(x) \mid x \in \text{dom } f\}.$$

[2] The "graph," as used in this discussion, is a diagram or "plot" of the function. If the domain of a function is a large finite set, a complete graph may be very extensive. If the domain is an infinite set, a graph can still be an informative schematic picture of the function.

Let us apply these conventions to a particular instance. Suppose that f is a function whose domain is the set of numbers in the closed interval from -1 to 1 and whose rule is: To each number in the domain, assign the positive square root of 1 minus its square. Since we deduce from the structure of the real number system that this is a unique assignment, the rule is adequate, and the domain is defined. This is very clear, but the student of mathematics usually finds

$$f: f(x) = \sqrt{1-x^2}, \quad -1 \leqslant x \leqslant 1$$

more concise and hence clearer. Not always is it possible, however, to convey a function from your mind to your reader with a single line of familiar symbols, as we soon shall see. Simply remember that if someone asserts, "Define a function f by . . . ," you must demand that his assertion contain *both* the domain and the rule before you give him more of your valuable attention.

Frequently, a statement is abbreviated when the reader can accurately supply what is omitted. We might write, "Consider the function $f(x) = \sqrt{1-x^2}$," wherein the function symbol f does not appear by itself, nor is the domain specified. This statement means that a function f is present whose domain is the set of all real numbers for which the formula $\sqrt{1-x^2}$ represents a real number and whose rule assigns to x in this domain $f(x) = \sqrt{1-x^2}$ of the range. This is the same function as the one given in the preceding paragraph.

If the function we have in mind is f, where

$$f: f(x) = \sqrt{1-x^2}, \quad 0 < x \leqslant 1,$$

there is no shortcut to the full description (notice the domain).

In practice we sometimes encounter two descriptions of functions which appear quite different, yet we suspect that they may describe the same function. In order to settle such a suspicion we are led to the following.

CRITERION 2.2 Two functions f and g are *equal* in case,

(a) dom $f =$ dom g,
(b) and if $x = y$ in the common domain, then $f(x) = g(y)$.

For example, define f, g, and h by

$$\begin{aligned} f&: f(x) = \sqrt{1-x^2}, & 0 \leqslant x \leqslant 1 \\ g&: g(t) = \sqrt{1-t^2}, & 0 \leqslant t \leqslant 1 \\ h&: h(y) = \tfrac{1}{2}\sqrt{4-4y^2}, & 0 \leqslant y \leqslant 1. \end{aligned}$$

We call these descriptions *equivalent,* since they describe the same function; hence, $f = g = h$.

EXAMPLE 1 $f(x) = \sqrt{x^2 - 1}$.

In this case we assume that dom $f = (-\infty, -1] \cup [1, \infty)$ is meant.

EXAMPLE 2
$$\begin{aligned} f(x) &= 1 - 2x, & 0 \leqslant x \leqslant 1 \\ &= 4x\quad, & 1 < x \leqslant 2. \end{aligned}$$

In this case dom $f = [0,2]$ and the image $f(x)$ of a domain number x is assigned by the first formula, if $x \leqslant 1$, by the second formula, if $x > 1$. It takes two lines to write out the rule.

EXAMPLE 3 (Post Office function)
First description: If x is the weight of a first class parcel in ounces, then $f(x)$ is the postage in cents necessary to mail it.
Second description (as of the time this is being written, postage is 6 cents per full ounce up to 70 pounds):

$$\begin{aligned} f(x) &= 6\quad, & 0 < x \leqslant 1 \\ &= 12\quad, & 1 < x \leqslant 2 \\ &= 18\quad, & 2 < x \leqslant 3 \\ &\ \ \vdots & \\ &= 6720, & 1119 < x \leqslant 1120. \end{aligned}$$

Third description:

$$f(x) = 6k, \quad k - 1 < x \leqslant k, \quad k = 1,2,3, \ldots, 1120.$$

Fourth description:

$$\begin{aligned} f(x) &= 6\quad, & 0 < x \leqslant 1 \\ f(x) &= f(x-1) + 6, & 1 < x \leqslant 1120. \end{aligned}$$

These descriptions are listed in decreasing order of practical clarity and increasing order of mathematical brevity. What do you think about this ranking?

2.3 PROBLEMS

1. For each of the following functions, state the domain and write out in words the rule of the function.
   (a) $f(x) = 1 + 2x$
   Answer: Domain is all numbers. Rule is: To each number x in domain assign the number obtained by adding 1 to the product of 2 and x.

(b) $f(x) = 1 + 2x, \quad 0 \leq x < 3$
(c) $f(x) = 1 + 2/x$

Answer: Domain is all numbers but 0. Rule is: To each number x in domain assign the number obtained by adding 1 to the quotient of 2 by x.

(d) $g(x) = 1 + x^2 + x^3$

(e) $g(x) = \dfrac{1 + x^2}{1 - x^3}$

(f) $h(x) = \begin{cases} 1 + 2x, & 0 \leq x \leq 10 \\ -4x, & 10 < x \leq 20 \end{cases}$

(g) $h(x) = \begin{cases} x + 1, & 0 \leq x \leq 1 \\ x - 1, & 3 \leq x \leq 4 \\ 5, & x = 9 \end{cases}$

(h) $\phi(x) = x + 1$
(i) $f(t) = \sqrt{1 + t}$

Answer: Domain is the interval $[-1,\infty)$. Rule is: To each number t assign the positive square root of the number obtained by adding 1 to t.

(j) $f(t) = \sqrt{1 + t^2}$
(k) $f(t) = \sqrt{1 - t^2}$
(l) $g(s) = \sqrt{1 - s^2}, \quad 1 \leq s$

(m) $g(y) = \begin{cases} y, & -1 \leq y \leq 0 \\ |y|, & 0 < y \leq 1 \end{cases}$

(n) $\phi(w) = \begin{cases} 1, & 1 \leq w < 2 \\ 2, & 3 < w \leq 4 \\ 3, & 6 < w < 7 \end{cases}$

2. Specify each of the following functions in symbols.
   (a) Domain is $1 \leq x \leq 2$. Rule is: For each x in domain add 3 to x and divide result by 2.

   Answer: $f(x) = (x + 3)/2, \quad 1 \leq x \leq 2$

   (b) Domain is $0 \leq x$. Rule is: For each x in domain take cube root of sum of x and twice the square of x.
   (c) Domain is $0 \leq y \leq 2$. Rule is: For each y in domain take midpoint of segment joining y to -1.
   (d) Rule is: For each s take half the area of a geometric square of side s.

   Answer: $f(s) = \frac{1}{2}s^2, \quad 0 < s$

   (e) Rule is: For each s take the average of the volume of a geometric cube of side twice s and the area of circle of radius one-half s.
   (f) Domain is $0 \leq t \leq 1$ together with $2 < t < 3$. Rule is: The number assigned to any t in the domain is $2t$.
   (g) Domain is $-2 \leq x \leq -1$ together with $1 \leq x \leq 2$. Rule is:

Number assigned to any x in first interval is $1 - x$, in second interval is 5.

3. For each of the functions in Problem 1 find the value, if any, of the function at
   (a) 0
   Answer: (a) 1, (d) 1
   (b) 1
   (c) -1
   Answer: (a) -1, (d) 1
   (d) 3
   (e) p
   (f) $p + q$
   Answer: (a) $1 + 2p + 2q$
4. For each of the functions of Problem 1
   (a) find the value of the function at 1 plus the value of the function at 3 whenever 1 and 3 are both in dom f;
   (b) find the value of the function at 4 whenever 4 is in dom f;
   (c) compare your answers to (a) and (b) to see when they are equal.
5. For which of the functions in Problem 1 is the number 0 in the range of the function?
6. For each of the following functions find all the numbers in dom f at which the value of the function is 2.
   (a) $f(x) = 1 + 2x$
   Answer: $\frac{1}{2}$
   (b) $f(x) = 2 - x^2$
   (c) $f(x) = \sqrt{1 - x^2}$
   Answer: none
   (d) $f(x) = \sqrt{9 - x^2}$
   (e) $f(x) = 1 - 2x, \quad 0 \leqslant x \leqslant 1$
   (f) $f(x) = x^3, \quad -1 \leqslant x \leqslant 1$
7. Prove that $f = g$ in each of the following cases.
   (a) $f(x) = \dfrac{|x|}{x^2}, \quad x \neq 0; \quad g(y) = \dfrac{1}{|y|}, \quad y \neq 0$
   (b) $f(x) = 2 \sin x \cos x; \quad g(y) = \sin 2y$
   (c) $f(x) = \log |x|; \quad g(y) = \frac{1}{2} \log y^2$
8. Show why $f \neq g$ when
$$f(x) = \frac{x^2 - 1}{x + 1}, \quad g(y) = y - 1, \quad y \text{ is any real number.}$$
9. Prove that the third and fourth descriptions of the Post Office function are equivalent.
10. Prove that f is a function if, and only if, x,y in dom f with $x = y$ implies $f(x) = f(y)$.

2.4 TO VISUALIZE A FUNCTION

As a mathematical entity, a function is an object. Mathematicians, in common with artists, attempt to realize an abstract concept in visual media of several kinds. For functions, this helps to suggest and direct ways to examine them, and in return, it helps to apply results of such an examination to the study of nonmathematical media.

One may think of a function as a supply of numbers serving as the input to a machine, also a part of the function, which cranks out the image of each number from the input. Figure 2.3 illustrates this concept. This is a rudimentary computer.

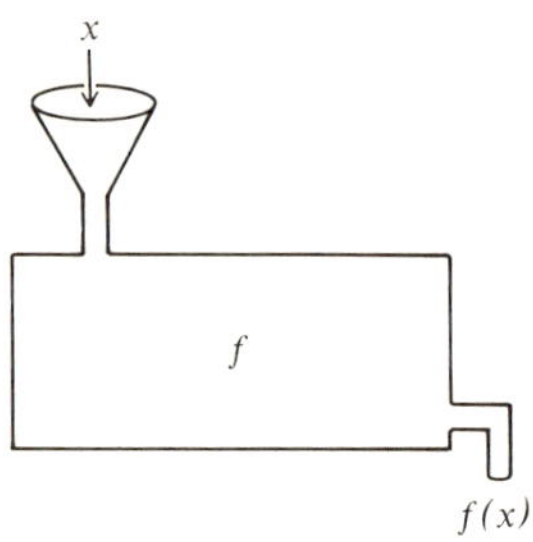

Figure 2.3

As another image of a function, think of an idealized table of numbers. In one column are all the numbers of the domain, and opposite each of these is its corresponding range number. This view leads to a collection of ordered pairs of numbers, the first number of a pair being an element of the domain; the second, its image in the function. This collection of ordered pairs of numbers is called the *graph* of the function.[3] Some authors define a function to be just such a collection of ordered pairs of numbers, no two pairs having the same first element. The set of ordered pairs is a very useful model for the function as we have defined it.

In fact, the familiar geometric graph, or plot, of a function, which can often be sketched on a drawing board reasonably faithfully, is an especially useful representation of the function. If we use the number pair $(x,f(x))$ as coordinates of a point P in the plane, then a function f gives rise to the point set[4]

$$\{P|P(x,f(x)), \quad x \in \operatorname{dom} f\},$$

[3] This graph is a set. It is different from the "graph" discussed in Section 2.2, which is a plot drawn on paper or a blackboard.

[4] The symbol $P(x,f(x))$ can be read "the geometric point P whose coordinates are $(x,f(x))$." In this discussion we refer to rectangular coordinates only.

the geometric graph of f. This point set is usually called a graph of f whenever the distinction between the two kinds of "graphs" is clear. Notice that if $P_1(x_1,f(x_1)) \neq P_2(x_2,f(x_2))$, then necessarily $x_1 \neq x_2$, for otherwise $(x_1 = x_2)$, the rule for f, would insist that $f(x_1) = f(x_2)$ and $P_1 = P_2$. Thus, no two points on the graph of f can lie on the same vertical line, and the arrangement of points on a graph of f *cannot* include such a configuration as is illustrated in Figure 2.4.

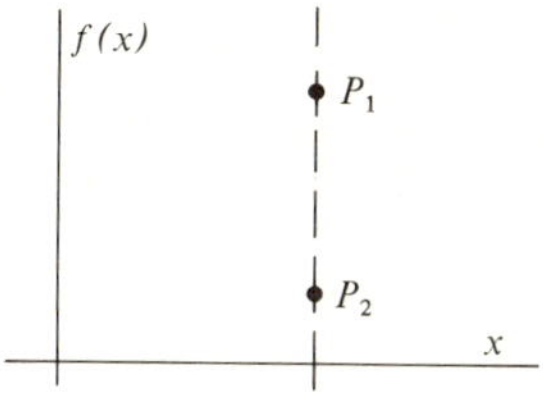

Figure 2.4

Some standard curves from analytic geometry can be used for geometric graphs of functions. Just for consistency, let x represent a number in the domain. For example, the line

$$Ax + By + C = 0, \quad x \text{ real}, \quad B \neq 0$$

is the graph of the function

$$f\colon f(x) = -\frac{A}{B}x - \frac{C}{B}, \quad x \text{ real}, \quad B \neq 0.$$

See Figure 2.5.

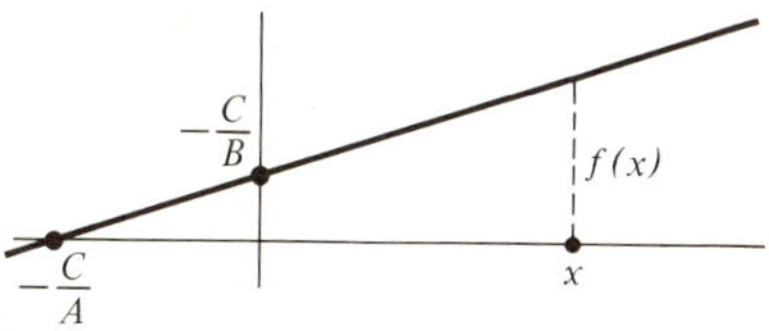

Figure 2.5

The line

$$Ax + By + C = 0, \quad A \neq 0, \quad B = 0$$

could be used for the graph of a function f only if the domain of that function was the singleton set dom $f = \{-C/A\}$. However, since both points $P_1(-C/A,y_1)$ and $P_2(-C/A,y_2)$, $y_1 \neq y_2$ are on this line, no segment of this line could be the graph of f, and this line is not of much use as a graph of a function. Refer to Figure 2.6.

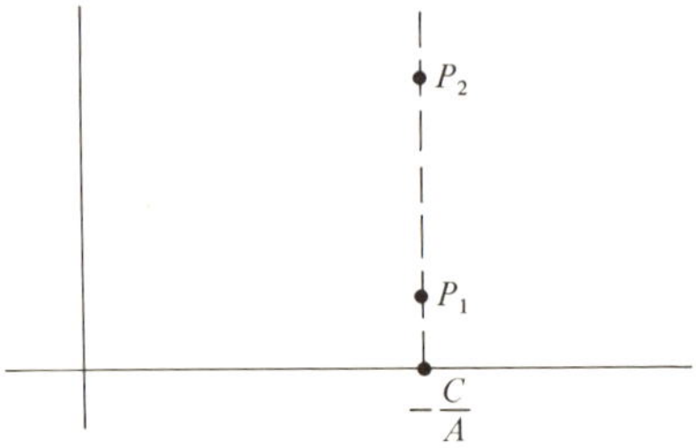

Figure 2.6

The parabola $x^2 = 4py$, $p > 0$ is the graph of the function

$$f\colon f(x) = \frac{1}{4p}x^2, \quad x \text{ real.}$$

See Figure 2.7.

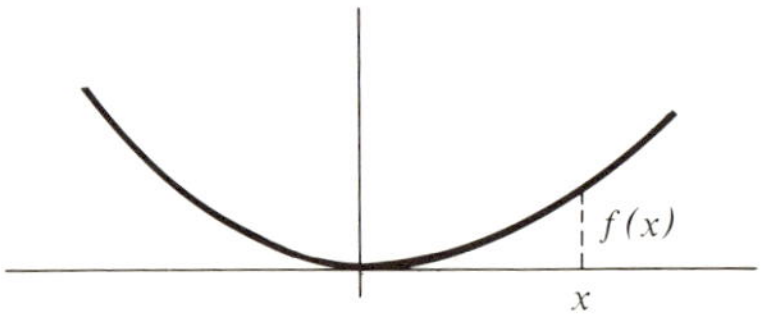

Figure 2.7

If the parabola is rotated 90° so that its equation is $y^2 = 4px$, $p > 0$, then a portion of it can be used for the graph of such a function as

$$\begin{aligned} f\colon f(x) &= 2\sqrt{px}\ , \quad x \geqslant 0, \quad x \text{ rational,} \\ f(x) &= -2\sqrt{px}, \quad x > 0, \quad x \text{ irrational,} \end{aligned}$$

although anything close to a faithful graphical representation of this function is beyond the skill of a draftsman. See Figure 2.8.

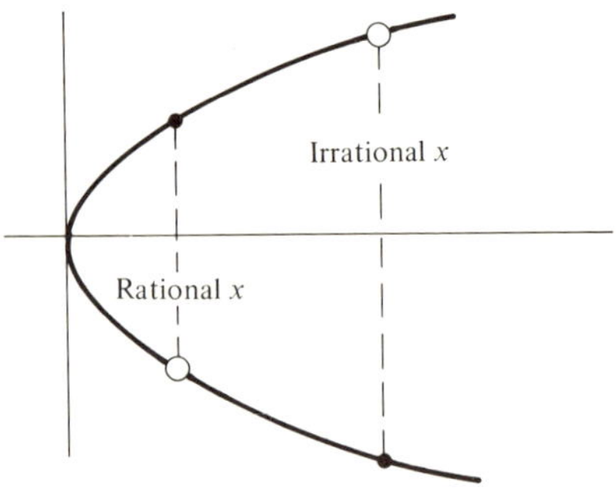

Figure 2.8

In a similar vein, the circle $x^2 + y^2 = 1$ has portions which are useful as graphs of such functions as

f: $f(x) = \sqrt{1 - x^2}$, $-1 \leqslant x \leqslant 1$ (upper semicircle)
g: $g(x) = -\sqrt{1 - x^2}$, $-1 < x < 1$ (lower semicircle with end points removed)

h: $h(x) = \dfrac{|x|}{x}\sqrt{1 - x^2}$, $x \in [-1,1]$, $x \neq 0$

$h(0) = 0$ (quarter circles in first and third quadrants, each with one end point removed, plus the center).

See Figures 2.9 and 2.10.

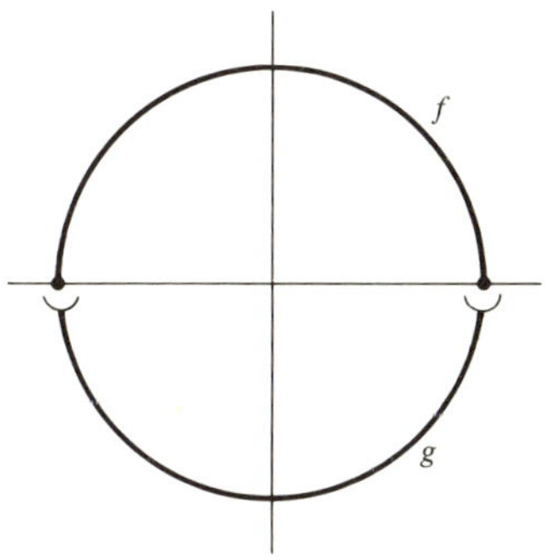

Figure 2.9

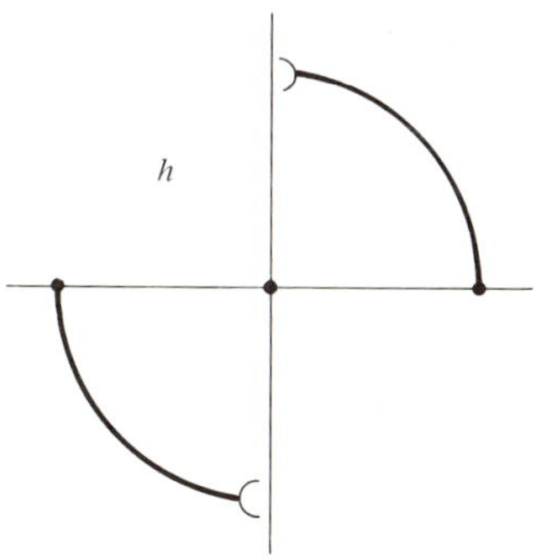

Figure 2.10

2.4 PROBLEMS

1. Sketch a graph of each of the following functions.
   (a) $f(x) = 2x$ (b) $f(x) = 2x + 1$
   (c) $f(x) = 2x$, $0 \leqslant x \leqslant 1$ (d) $f(x) = x$
   (e) $f(x) = x/2$ (f) $f(x) = x/10$

(g) $f(x) = 0$

(h) $g(x) = 1$

(i) $g(x) = |x|$

(j) $g(x) = |x| + 1$

(k) $g(x) = |x + 1|$

(l) $h(x) = \sqrt{4 - x^2}$

(m) $h(x) = -\sqrt{4 - x^2}$

(n) $h(x) = 1 + \sqrt{4 - x^2}$

(o) $f(t) = t^2$

(p) $f(t) = 1 - 2t^2$

(q) $f(y) = y^3$

(r) $f(y) = 1/y$

(s) $f(s) = \begin{cases} s, & 0 \leqslant s \leqslant 1 \\ 2s - 1, & 1 < s \leqslant 2 \end{cases}$

(t) $\phi(z) = \begin{cases} z, & 0 \leqslant z \leqslant 1 \\ 1 - z, & 1 < z \leqslant 2 \end{cases}$

(u) $\phi(t) = \begin{cases} t, & 0 \leqslant t \leqslant 1 \\ 1, & 1 < t \leqslant 2 \end{cases}$

(v) $f(w) = \begin{cases} w, & 0 \leqslant w \leqslant 1 \\ 1, & 2 \leqslant w \leqslant 3 \end{cases}$

(w) $g(x) = \begin{cases} x, & 0 \leqslant x \leqslant 1 \\ 2x, & 2 < x \leqslant 3 \\ x/2, & 3 < x \leqslant 4 \end{cases}$

2. Describe the functions for which the following are the graphs:
   (a) the line through the origin with slope -2
   Answer: $f(x) = -2x$
   (b) the line through (1,2) with slope 3
   Answer: $f(x) = 3x - 1$
   (c) the line joining (1,2) and (p,q)
   (d) the upper half of the circle with center at (1,2) and radius 3
   (e) the segment joining $(-2,2)$ to $(2,-1)$
   Answer: $f(x) = 1 - x, \quad -1 \leqslant x \leqslant 2$
   (f) the part of the circle $x^2 + y^2 = 4$ which lies in the second quadrant
   (g) the segment joining (0,0) to (1,1) and the segment joining (1,1) to (2,0)
   (h) the segment joining (0,0) to (1,1) and the segment joining (1,1) to (2,1).
3. Construct four different functions, each of whose graphs are portions of the ellipse $b^2x^2 + a^2y^2 = a^2b^2$.
4. Of all the hyperbolas, find those which, in their entirety, can be graphs of functions.

2.5 SOME ORIGINS OF FUNCTIONS

Relationships between measurable quantities arise in many fields, and often they can be described by real-valued functions such as those we are illustrating here. To derive information about such a relationship we must find a function which corresponds to the relationship. There are no fixed rules for doing this, but with practice you can obtain some expertise and see how the theory of functions can be used in applications of mathematics. Many of the older textbooks included lists of story problems, and motivated the study of mathematics by emphasis on its applicability. In this book we assume that you are interested primarily in the mathematics itself, though we occasionally refer to its applications, as we are doing here.

The following problem list provides illustrations of how questions about relationships which can arise in nonmathematical areas can be rephrased in the language of functions, and, in some cases, answered.

2.5 PROBLEMS

1. The area of a circle is related to its radius. What is the function which expresses this relationship?
   Answer: $f(x) = \pi x^2$, $0 < x$
   Remark: Another way of stating the above problem is, "Find the area of a circle as a function of its radius." This terminology, while often used, lays stress on the value of the function, in this case the area, instead of on the function itself. While we shall use this terminology occasionally, remember that it is the function itself which is required.
2. Find the area of a semicircle as a function of its perimeter.

   Solution: The perimeter p of a semicircle is half the circumference of the circle plus the diameter, or $p = \pi r + 2r$, where r is the radius. Hence, $r = p/(\pi + 2)$, and the area of the semicircle is $\pi r^2/2 = \pi p^2/2(\pi + 2)^2$. Therefore, the function is given by

   $$f(x) = \frac{\pi x^2}{2(\pi + 2)^2}, \quad 0 < x.$$

3. Express the surface area of a cube as a function of its edge.
4. A taxi charges 20 cents for the first $\frac{1}{4}$ mile and 15 cents for each succeeding $\frac{1}{4}$ mile. What is the function which relates cost to distance traveled?
5. From a rectangular piece of paper 10 inches by 20 inches, equal squares are cut out at each corner and the edges are folded up to make a box, as in Figure 2.11. Find the volume of the box in terms of the size of the squares.

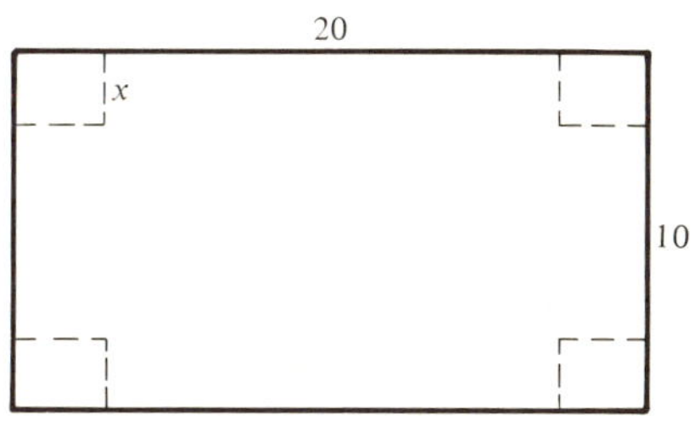

Figure 2.11

Solution: Denote the side of the squares by x. Then the base of the box will be $10 - 2x$ inches by $20 - 2x$ inches. The height will be x inches. Therefore, the volume will be $x(10 - 2x)(20 - 2x)$, and the function is given by $f(x) = x(10 - 2x)(20 - 2x)$, $0 < x < 5$.

6. A rectangular box with a cover is to have a square base. The volume is to be 10 cubic feet. Find the surface area of the box in terms of the side of the base.

   Solution: Let x represent the side of the base. Then x^2 is the area of the base. Since the volume is 10, the height must be $10/x^2$. Each side has area $x10/x^2 = 10/x$. There are four sides, a top, and a bottom, so that the total surface is $40/x + 2x^2$. The function is given by $f(x) = 40/x + 2x^2$, $0 < x$.

7. If the material for the base of the box of Problem 6 costs 10 cents per square foot, for the sides 20 cents, and for the top 5 cents, find the cost of the box as a function of the side of the base.
8. A rectangle is inscribed in a circle of radius R. Find its area as a function of the length of one side.
   Answer: $f(x) = x\sqrt{4R^2 - x^2}$, $0 < x < 2R$
9. A window has the shape of a rectangle surmounted by a semicircle. If the perimeter is 100 inches, find the function which relates the area of the window to the length of the base of the rectangle.
10. 1000 yards of fencing are to be used to enclose two plots, one a square and one a circle. Find the total area enclosed in terms of the side of the square.
11. The cost per hour for fuel to run a boat is proportional to the cube of the speed and is 10 dollars for a speed of 5 miles per hour. Other costs are 20 dollars per hour. Find the cost of going 10 miles as a function of the speed.
    Answer: $f(x) = \frac{4}{5}x^2 + \frac{2000}{x}, 0 < x$
12. The strength of a beam is proportional to the breadth and the square of the depth. Find the strength of beams which can be cut from a circular log as a function of their breadths.
13. A triangle has two of its vertices at the points (0,0) and (4,0). Its third vertex is on the graph of the equation $x^2y = 1$. Find the function which gives the area of the triangle in terms of the x-coordinate of the third vertex.
    Answer: $f(x) = 2/x^2$
14. A right circular cylinder is inscribed in a sphere of radius 1. Find its volume as a function of the radius of its base.
15. A tank has the form of a right circular cylinder with hemispherical ends. Its volume is A. Find its total length as a function of its radius.
16. A right circular cylinder is inscribed in a right circular cone of height

H and radius R. Find the function which relates the volume of the cylinder to the radius of its base.

17. A radio manufacturer spends 200 dollars a week for rent and upkeep of his plant. His production cost is 10 dollars per radio set. At a selling price of p dollars per set he can sell $200 - p$ sets per week. Find the function which expresses his weekly profit as a function of selling price.

Solution: His income is $(200 - p)\,p$ dollars and his total cost is $200 + 10(200 - p)$. Hence his profit is $210p - p^2 - 2200$. But his selling price must be a positive number (we assume he does not pay people to carry his sets away), so $p > 0$. Also, the number of sets he sells, $200 - p$, must be nonnegative, so $200 - p \geqslant 0$. Hence, $p \leqslant 200$, so p must be in the interval $(0,200]$. Finally, he can sell only a whole number of sets, so $200 - p$ and hence p itself must be an integer. Therefore, the domain of definition of our function consists of the integers in the interval $(0,200]$, and the rule is

$$f(p) = 210p - p^2 - 2200.$$

18. Use the information given in Problem 17, with the exception that the manufacturer can sell only $200 - 4p$ at a selling price of p.
19. Use the information given in Problem 18, with the exception that the manufacturer has other costs of $n^2/100$ dollars if he produces n radios per week.

2.6 MORE EXAMPLES OF FUNCTIONS

The definition of what a function is to be, such as Definition 2.1, has evolved from studies of relations between sets of measurements. The examples in the problem list of the last section illustrate this kind of study. Our point of view is, in one sense, the reverse of this. We decide what a function is to be in essence and then construct and classify many functions from which an applied mathematician may select those which are useful to him.[5]

The quantity of functions created by Definition 2.1 beggars description. We may get some idea of the immense number of functions which exist by the following scheme for listing, relatively, just a few of them. Let a function be described by its graph (set of ordered pairs). How many functions are there whose domains are subsets of the set of only two num-

[5] In response to a question, a contemporary scientist, J. Robert Oppenheimer, remarked that physicists have invented their own mathematical theories. But he added that subsequently these theories have usually been discovered to already be in the mathematical literature.

bers, $\{0,1\}$ and whose ranges may be selected from the same simple subsets? In order to simplify the enumeration we temporarily allow a domain to be empty. A list of these functions, by means of their graphs, is

$$f_1 = \varnothing, \quad f_2 = \{00\}, \quad f_3 = \{01\}, \quad f_4 = \{10\}, \quad f_5 = \{11\},$$
$$f_6 = \{00\ 10\}, \quad f_7 = \{00\ 11\}, \quad f_8 = \{01\ 10\}, \quad f_9 = \{01\ 11\}.$$

There are nine of them, two of whose geometrical graphs are shown in Figure 2.12.

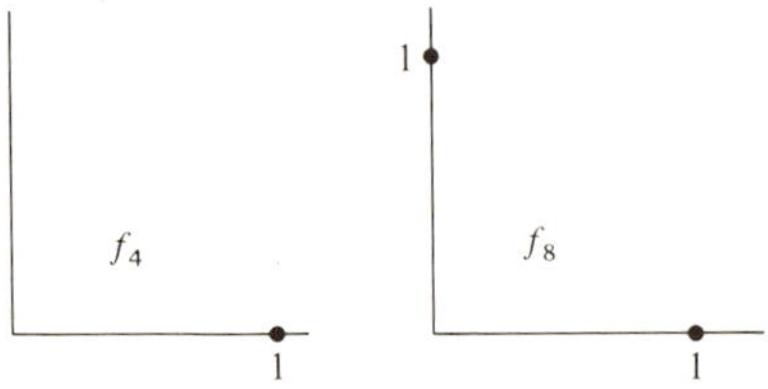

Figure 2.12

You can verify that if the domains and ranges are drawn from the set of numbers $\{0,1,2, \ldots, n-1\}$ in this way, the number of constructable functions is $(n+1)^n$. For the above case $n = 2$; therefore, the number of functions is $3^2 = 9$. If the basic set contained only nine numbers, there would be 10^9 functions—already a billion of them! In many, if not most, applications of functions more than nine measurements are made, so we see the wealth of options from which to choose a descriptive function. In our study the basic set from which domains and ranges are selected is the uncountable set of real numbers, and the collection of functions we will attempt to classify is certainly nontrivial, to put it mildly.[6]

If you glance ahead you will notice that after some technical work we will classify our functions according to certain properties they may have in common. Some functions are bounded; some are continuous, differentiable, or integrable. To get ready for this we pause again to look at a few samples, this time paying attention to breaks and corners on their graphs (plots).

SAMPLE 1 $\quad f\colon f(x) = \dfrac{|x|}{x}, \quad x \neq 0$

$$f(0) = 1$$

The domain of f is the set of all real numbers, and each point on the graph is on one of two horizontal rays (see Figure 2.13). The function is of

[6] For infinite sets the "number of elements" loses its meaning. In books on set theory *cardinality* is used to characterize the size of a set, and the cardinality of the set of functions is greater than the cardinality of the basic domain set as we used it.

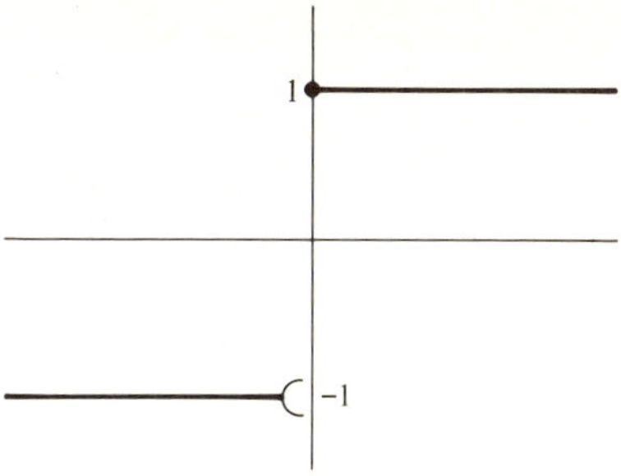

Figure 2.13

little immediate interest except at $x = 0$, where the graph separates from itself. This fact can be stated in the following way, which will be useful later on. We can choose an interval I in the range, say, with length $\frac{1}{2}$, which contains 1 at its center,[7] such that for any number x' in the domain there will be another number x closer to 0 whose image, $f(x)$, is outside of I. See Figure 2.14. More specifically, for any $\delta > 0$ there is an x with $|x| < \delta$, so that $|f(x) - 1| > \frac{1}{4}$. An easy formula for such an x is $x = -\delta/2$, so that $|x| = \delta/2 < \delta$ and $|f(x) - 1| = |-1 - 1| = 2 > \frac{1}{4}$.

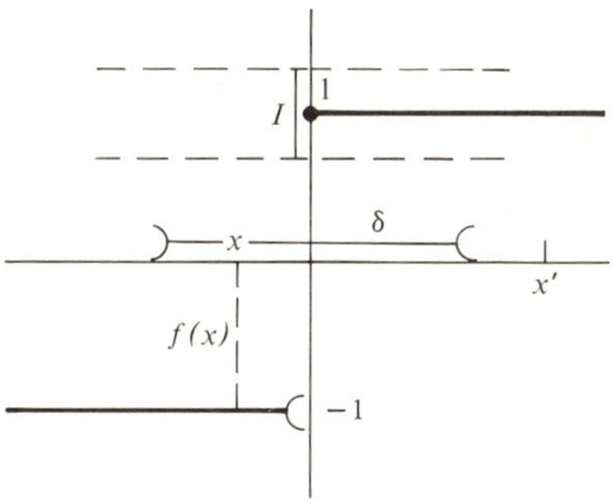

Figure 2.14

SAMPLE 2 $\quad f\colon f(x) = 1 \quad x$ rational
$\qquad\qquad\qquad f(x) = -1 \quad x$ irrational

A geometric graph of this function can only poorly be suggested by a diagram, as the one in Figure 2.15. This graph separates from itself at every point. If we apply a criterion such as was used in Sample 1, this property of the function becomes more precise. Choose an interval I of

[7] Any such interval with length less than 4 would suffice to identify the break in the graph, since it would not cover the jump from -1 to 1.

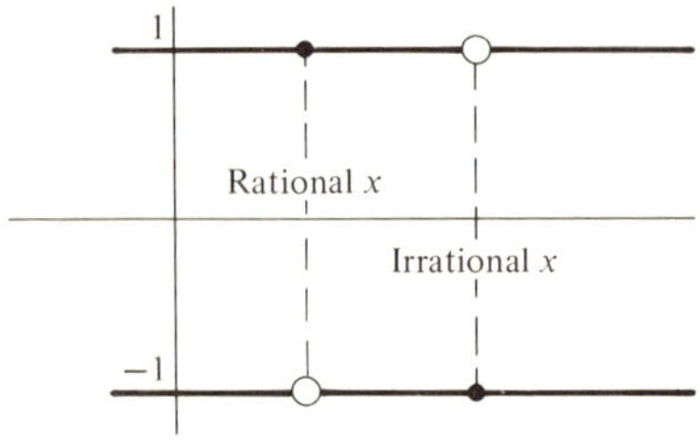

Figure 2.15

length $\frac{1}{2}$ centered at 1 in the range, and any rational number x_0 in the domain. Let $\delta > 0$ be a nonzero number, say rational. Then $x = x_0 + \delta/\sqrt{2}$ is irrational,[8] with

$$|x - x_0| = \frac{\delta}{\sqrt{2}} < \delta.$$

Since $f(x) = -1$,

$$|f(x) - 1| = |-1 - 1| = 2 > \frac{1}{4}.$$

See Figure 2.16.

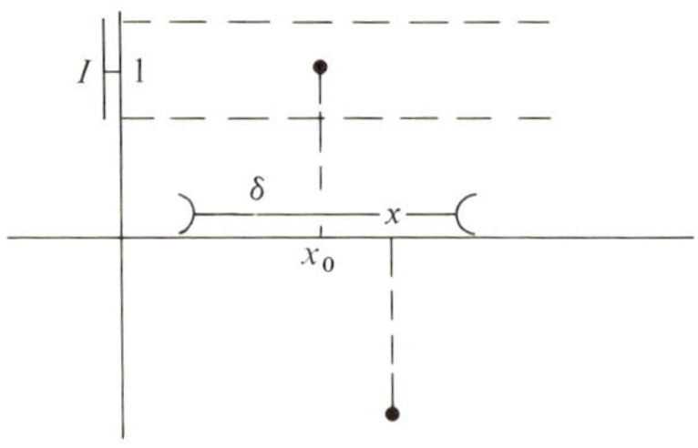

Figure 2.16

This is an algebraic way of proving that however one restricts domain numbers to be close to x_0, there will always be such domain numbers whose images are distant from $f(x_0)$, by more than $\frac{1}{4}$ in fact.

SAMPLE 3 $\quad f: f(x) = x \sin \frac{\pi}{x}, \quad x \neq 0$

$$f(0) = 0$$

[8] The number $\sqrt{2}$ is chosen only because it provides a number $\delta/\sqrt{2}$ which is less than δ. Any other irrational denominator greater than 1 would do.

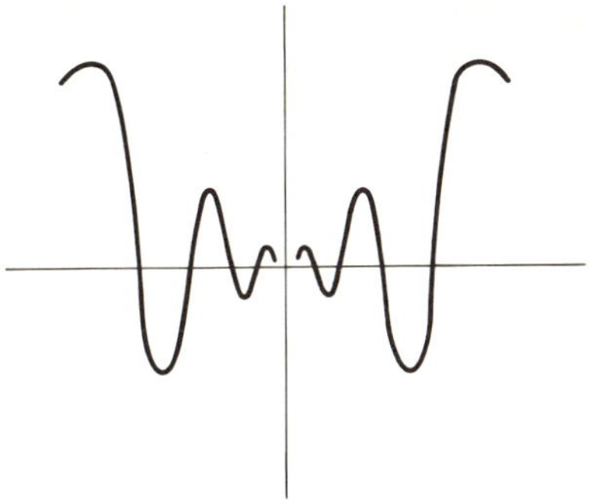

Figure 2.17

The sketch of the graph of f is shown in Figure 2.17. One would expect this graph not to separate from itself at any point in the sense that the preceding two samples did, unless something unusual happened at $x = 0$, where the branches of the curve are jammed together and impossible to draw. Let us use the kind of analysis we employed in the other two samples to verify that the curve does not separate from itself at the origin. To do this we will demonstrate that for whatever interval, large or small, that we center on 0 in the range, there can be found an interval centered on 0 in the domain, all of whose points have their images in that range interval. Let the range interval I have length 2ϵ, where $\epsilon > 0$. Take $\delta = \frac{1}{2}\epsilon$. Then if x is any number such that $0 < |x - 0\ | < \delta$, it follows that

$$|f(x) - 0| = |x|\ |\sin \pi/x| \leqslant |x| < \delta$$
$$= \frac{1}{2}\epsilon < \epsilon.$$

Study Figure 2.18 to keep this argument straight. Remember that $0 < |x - 0| < \delta$ means that $x \neq 0$ and $-\delta < x < \delta$, and that $|f(x) - 0| < \epsilon$ means that $-\epsilon < f(x) < \epsilon$.

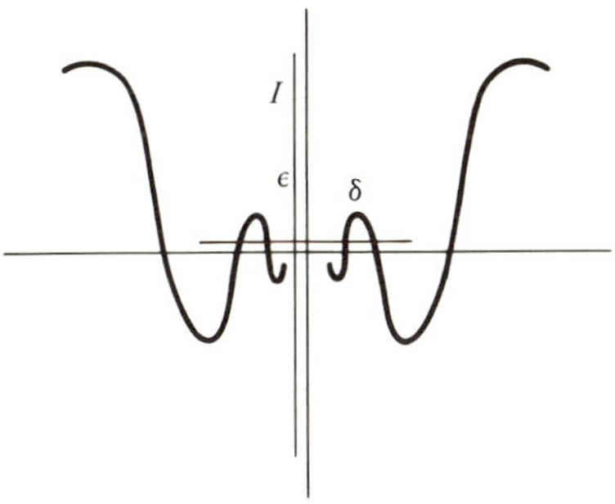

Figure 2.18

In this instance we have not really proved anything about the geometric graph of f at 0 because no matter what scale we use on the horizontal axis there will always remain intervals containing 0 that we cannot see. But we have proved something about the function f; namely, whatever range interval I, containing $f(0)$, is taken, there will be a domain interval containing 0 that maps into I.

SAMPLE 4 A gram of ice at $-10°$ C is put into a warm oven. Let the function f have the rule: For each x the image $f(x)$ is the temperature of the water after it has absorbed x calories of heat. The domain of f is a sufficiently large interval of numbers to measure the calories absorbed in the experiment, say [0,800]. A graph of this function is shown in Figure 2.19.

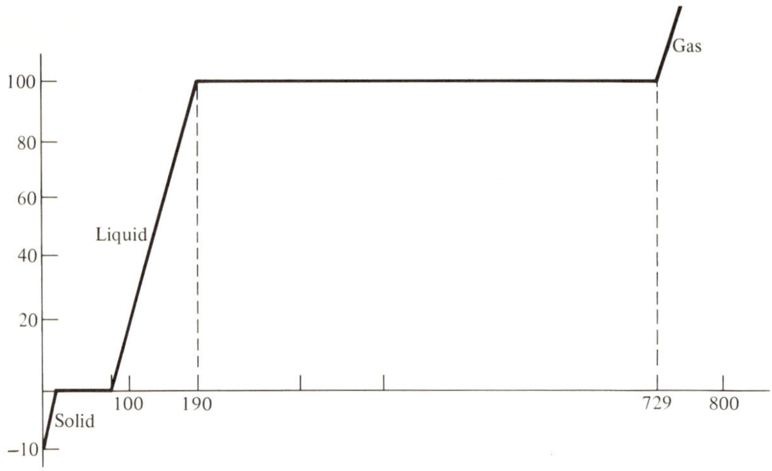

Figure 2.19

As in Sample 4, we could show that this graph does not separate from itself anywhere; it has no breaks. But we observe that it has corners, and we look for a way to write this out in the language of functions. You should verify that a description of f is

$$\begin{aligned} f\colon f(x) &= x - 10 & 0 &\leq x < 10 \\ f(x) &= 0 & 10 &\leq x < 90 \\ f(x) &= x - 90 & 90 &\leq x < 190 \\ f(x) &= 100 & 190 &\leq x < 729 \\ f(x) &= x - 629 & 729 &\leq x \leq 800. \end{aligned}$$

At a corner the graph appears suddenly to change direction.

In order to detect a corner with algebraic analysis we borrow from railroad engineers a device which they use to measure the grade of a roadbed. Figure 2.20 depicts a portion of the graph of a function f and a point P on the graph. Let P have the coordinates $(x, f(x))$, and choose another point Q on the graph with coordinates $(x + h, f(x + h))$. Then the *slope* of the line segment joining P and Q is

$$\frac{f(x+h)-f(x)}{h}, \quad h \neq 0.$$

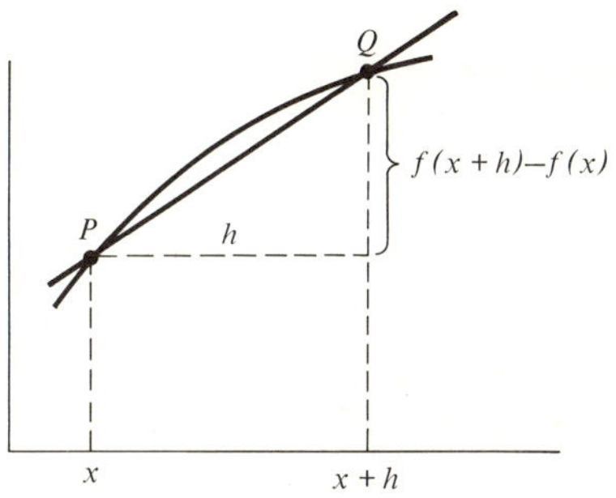

Figure 2.20

The engineers call this the *grade* of $\overline{PQ}$ when it is nonnegative. For each choice of $h \neq 0$, so that $x + h$ is in dom f, a unique slope is thus determined, and this rule defines the *slope function* for f at x.

$$h \longrightarrow \frac{f(x+h)-f(x)}{h}, \quad h \neq 0, \quad x+h \in \operatorname{dom} f.$$

A name for the slope function should include designations of both the function f and the number x from which it is derived. Such a name can be ${}_f\sigma_x$, and then the slope function for f at x is described by

$${}_f\sigma_x(h) = \frac{f(x+h)-f(x)}{h}, \quad h \neq 0, \quad x+h \in \operatorname{dom} f.$$

Applied to the Sample 4, one such slope function is

$$\begin{aligned} {}_f\sigma_{190}(h) &= \frac{1}{h}\,[f(190+h)-f(190)] \\ &= \frac{1}{h}\,[f(190+h)-100], \quad h \neq 0, \quad -190 \leqslant h \leqslant 610. \end{aligned}$$

Specifically, in the interval $-100 \leqslant h \leqslant 539$, $h \neq 0$, this slope function has the formulas

$$\begin{aligned} {}_f\sigma_{190}(h) &= 1, \quad -100 \leqslant h < 0 \\ &= 0, \quad 0 < h \leqslant 539. \end{aligned}$$

Thus, at $x = 190$, where the graph of f in Figure 2.21 exhibits a corner, the graph of the corresponding slope function would have a break, a jump of 1 at $h = 0$. We will see the slope function again soon.

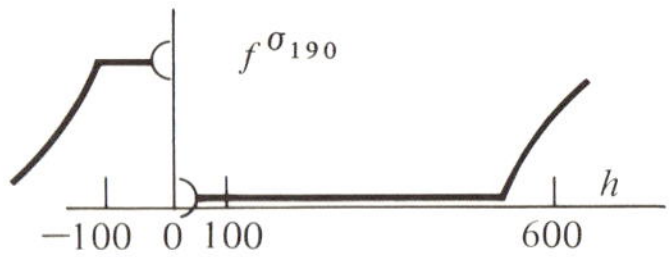

Figure 2.21

SAMPLE 5 On a certain quiz a class of 25 students received the grades

6	2	9	7	10
0	9	6	5	1
7	8	5	6	4
6	9	6	10	10
8	3	10	1	9

The cumulative frequency function for this list of numbers has a domain which is the interval [0,10] (from the smallest to the largest number in the list), and the rule: $f(x)$ is the number of numbers in the list not exceeding x. Since four students got grades of 2 or less, $f(2) = 4$. The median grade is the smallest domain number whose image is at the middle of the range (see Figure 2.22). If there had been an even number of students, the median grade would have been defined slightly differently. Statisticians find this graph useful, even though it has many breaks.

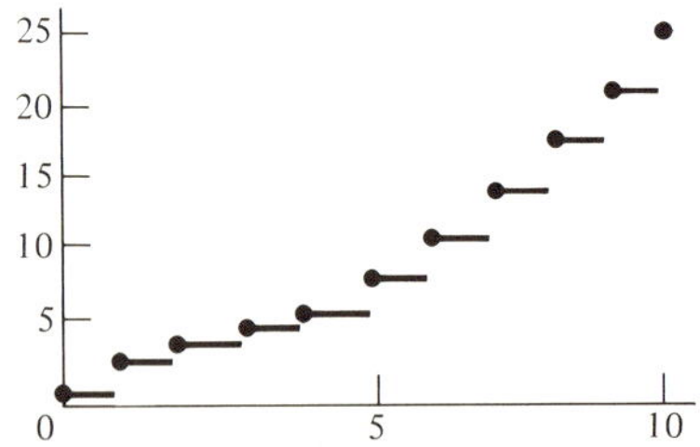

Figure 2.22

2.6 PROBLEMS

1. Let the domains and ranges of certain functions be subsets of $\{0,1,2\}$. How many such functions are there? Write out and plot geometric graphs of four (nontrivial) such functions.

2. Draw the graphs of the following.
   (a) $f(x) = |x|/x$
   (b) $f(x) = |x \sin(\pi/x)|$
   (c) $f(x) = x|\sin(\pi/x)|$
   (d) $f(x) = x^2 \sin(\pi/x)$
3. By the greatest integer in x, denoted by $[x]$, we mean the largest integer n such that $n \leqslant x$. Thus $[x] \leqslant x < [x] + 1$. Draw the graphs of the following.
   (a) $f(x) = [x]$
   (b) $f(x) = 2[x]$
   (c) $f(x) = [2x]$
4. Draw the graph of the function, defined in Section 2.3, giving the postage required to mail a letter.
5. Draw the graph of f, where f is defined by

$$f(x) = 3 + 3[x], \quad 0 < x < 1120.$$

   Is this the same function as in Problem 4?
6. Draw the graph of

$$f(x) = \begin{cases} 0, & 0 \leqslant x < 3 \\ 1, & 3 \leqslant x < 6 \\ 2, & 6 \leqslant x < 9. \end{cases}$$

   Write an equation for $f(x)$ in terms of the greatest integer function.
7. Draw the graph of the function defined as in Sample 5 for the following collection of numbers.

16	9	20	12
20	16	11	10
7	5	4	11
16	16	0	13
8	10	19	18
2	7	2	6
19	5	8	15
18	6	13	8
19	10	4	3
3	11	17	20

8. Let the domains of certain functions each be the set of natural numbers 0,1,2, ... Let the ranges of these functions be subsets of the ten digits 0,1,2, . . . ,9. Then an infinite sequence describes such a function. For example, if $f(0) = 1$, $f(1) = 9$, $f(2) = 8$, and $f(n) = f(n-3)$ for $n = 3,4,5, \ldots$, then the sequence description is repeating, as 198198 $\cdots$. There are "more" of these functions than there are natural numbers in the sense that there is no possible one-

to-one correspondence between the set of these functions and the set of natural numbers. To see why, do the following:

(a) Assume that there is such a correspondence, that is:

$$\begin{aligned} f_0 &\longrightarrow 198198 \cdots \\ f_1 &\longrightarrow 401826 \cdots \\ f_2 &\longrightarrow 333333 \cdots \\ &\vdots \end{aligned}$$

in which all the functions are listed.

(b) Define one of these functions f' as follows:

$$\begin{aligned} f'(n) &= f_n(n) + 1, \quad f(n) \neq 9 \\ &= f_n(n) - 1, \quad f(n) = 9 \end{aligned}$$

and show that $f' = f_j$ for no natural number j.

(c) Conclude that f' is not in the list of (a) and that the set of functions cannot be enumerated.

9. Let f be the function given by

$$\begin{aligned} f(x) &= x \qquad\;\, , \quad x < 0 \\ &= x^2 + 1, \quad x > 0. \end{aligned}$$

Prove that if $\epsilon = \frac{1}{2}$, then whatever $\delta > 0$ is chosen, there is a number x for which $0 < |x| < \delta$ and $|f(x) - 1| > \epsilon$. Explain how this analysis shows a break in the graph of f at 0.

10. For the function f of Problem 9 prove that whatever $\epsilon > 0$ is chosen there will always be a corresponding $\delta > 0$ so that if the number x satisfies $|x + 1| < \delta$, then necessarily $|f(x) + 1| < \epsilon$. Explain how this analysis shows that there is no break in the graph of f at -1.

11. Let f be the function given by

$$f(x) = \sin \frac{1}{x}, \quad x \neq 0.$$

Show that the graph of f has a break at 0.

12. Let f be the function given by

$$\begin{aligned} f(x) &= x^2 \sin \frac{1}{x}, \quad x \neq 0 \\ &= 0 \qquad\qquad , \quad x = 0. \end{aligned}$$

Show that the graph of f has no break at 0.

13. Complete the description of the slope function for the function f of Sample 4 at 190.

14. Describe ${}_f\sigma_{90}$, ${}_f\sigma_{100}$, and ${}_f\sigma_{500}$ for the function f of Sample 4. Sketch a graph of each of these slope functions near 0.
15. Let f be the function with domain [0,1] and the rule

$$\begin{aligned} &f(0) = 1 \\ &f(x) = 0, \quad x \text{ irrational} \\ &f(x) = \frac{1}{q}, \quad x = \frac{p}{q}, \quad \text{g.c.d. } (p,q) = 1 \\ &f(1) = 1. \end{aligned}$$

   (a) Prove that for any $\epsilon > 0$ there are only a finite number of numbers $x \in \operatorname{dom} f$ for which $f(x) \geqslant \epsilon$.
   (b) Show that for any $\epsilon > 0$ there is a corresponding $\delta > 0$ so that if a is irrational and $|x - a| < \delta$, then $|f(x)| < \epsilon$.
   (c) Show that if a is rational, then there is a corresponding $\epsilon > 0$ so that for any $\delta > 0$ there is a number $x \in \operatorname{dom} f$ with $|x - a| < \delta$ for which $|f(x) - f(a)| \geqslant \epsilon$.

2.7 AN ARITHMETIC OF FUNCTIONS

The operations of addition and multiplication of numbers extend easily to corresponding operations in the class of functions. Let f and g be functions. If dom f and dom g have a number x in common, then the numbers $f(x)$ and $g(x)$ have a unique sum $f(x) + g(x)$ and a unique product $f(x)g(x)$. This observation justifies the next definition.

DEFINITION 2.2(a) If dom $f \cap$ dom $g \neq \varnothing$, then

$$\begin{aligned} f+g&: (f+g)(x) = f(x) + g(x), \quad \operatorname{dom}(f+g) = \operatorname{dom} f \cap \operatorname{dom} g \\ fg&: (fg)(x) = f(x)g(x) \quad , \quad \operatorname{dom}(fg) = \operatorname{dom} f \cap \operatorname{dom} g. \end{aligned}$$

Notice that whenever the sum and product functions exist, that is, whenever the intersection of the domains of the given functions is not empty, both the rule and the domain for each of the sum and the product are specified. One might visualize the operations of forming the sum of two functions as the operation performed by a machine such as is shown in Figure 2.23. The box D contains a duplicator that makes a copy of the number dropped into it, and the box + is an adder. If the adder is replaced by a multiplier, one obtains a product machine.

EXAMPLE Let f and g be the functions

$$\begin{aligned} &f: f(x) = \frac{1}{x} \quad , \quad x \neq 0 \\ &g: g(x) = x - 2, \quad 0 \leqslant x \leqslant 2. \end{aligned}$$

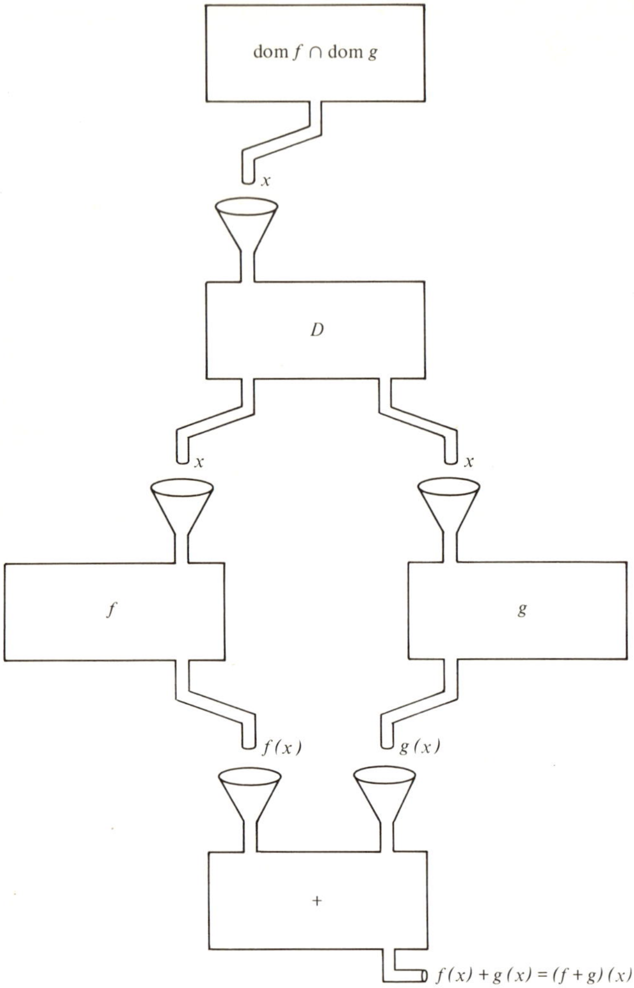

Figure 2.23

Then dom $f \cap$ dom $g = [x|0 < x \leqslant 2]$, and

$$f+g\colon (f+g)(x) = \frac{1}{x} + (x-2), \quad 0 < x \leqslant 2$$

$$fg\colon (fg)(x) = \frac{1}{x}(x-2), \quad 0 < x \leqslant 2.$$

An important special case of the product of two functions arises when the first factor is the constant function

$$f\colon f(x) = c, \quad c \text{ is a real number.} \quad \text{dom } f = \{\text{real numbers}\}.$$

If g is a function, then

$$fg\colon (fg)(x) = f(x)g(x) = cg(x),$$

where dom $fg =$ dom g. This can be viewed also as a scalar multiple of g, and is usually written simply as cg, that is,

$$cg\colon (cg)(x) = cg(x), \quad x \in \text{dom } g.$$

DEFINITION 2.2(b) If dom $f \cap$ dom $g \neq \varnothing$, then

$$f-g\colon (f-g)(x) = f(x) - g(x), \quad \text{dom}(f-g) = \text{dom } f \cap \text{dom } g$$
$$\frac{f}{g}\colon \frac{f}{g}(x) = \frac{f(x)}{g(x)}, \text{ dom } \frac{f}{g} = \text{dom } f \cap \text{dom } g - \{x|g(x) = 0\}.$$

These definitions of the difference and quotient of functions require some care in their application in case dom $f \neq$ dom g. Notice that if dom $f =$ dom g, then dom$(f-g)$ is this same set and

$$[(f-g)+g](x) = (f-g)(x) + g(x) = [f(x) - g(x)] + g(x) = f(x),$$

where dom$(f-g) \cap$ dom $g =$ dom f. This is consistent with the definition of the sum of two functions, and is simply formulated by $(f-g) + g = f$. However, if there is a number x in dom f which is not in dom g, then when we add g to $f-g$, the number x is not in the domain of the sum, and this sum is a different function than f. See Figure 2.24. In practice this trouble

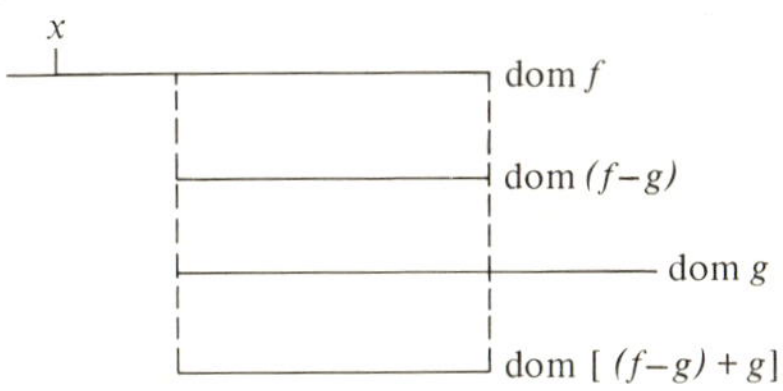

Figure 2.24

can often be avoided by replacing the functions to be added or multiplied by new functions which have equal domains. Suppose that dom $f \neq$ dom $f \cap$ dom g and/or dom $g \neq$ dom $f \cap$ dom g. Then, since the intersection is a subset of each of dom f and dom g, we can define the new functions f' and g' by

$$f'\colon f'(x) = f(x), \quad x \in \text{dom } f \cap \text{dom } g$$
$$g'\colon g'(x) = g(x), \quad x \in \text{dom } f \cap \text{dom } g$$

so that dom $f' =$ dom g'.

EXAMPLE Let f and g be the functions of the preceding example.

$$f-g\colon (f-g)(x)=\frac{1}{x}-(x-2),\quad 0<x\leq 2$$

$$\frac{f}{g}\colon \frac{f}{g}(x)=\frac{1}{x(x-2)},\quad 0<x<2$$

Note that $g(2)=0$.

In light of the above discussion, observe that

$$(f-g)+g\colon ((f-g)+g)(x)=\left[\frac{1}{x}-(x-2)\right]+(x-2)$$
$$=\frac{1}{x},\quad 0<x\leq 2,\quad \text{(compare with dom } f)$$

so that $(f-g)+g\neq f$. On the other hand, if we replace f by the new function

$$f\colon f(x)=\frac{1}{x},\quad 0<x\leq 2,$$

then $(f-g)+g=f$. Carry out this analysis for $(f/g)g$.

However awkward this attention to domains may appear, it does emphasize that a function has a richer structure than does a number in its range. After a few problems for practice, we will take advantage of this extra structure and use it to invent a new operation with functions which did not appear in the algebra of numbers. This new operation is a great laborsaving device in doing calculus, and we will return to it repeatedly.

2.7 PROBLEMS

1. Find $f+g$ and fg if
   (a) $f(x)=x;\quad g(x)=1$
   Answer: $(f+g)(x)=1+x$, all numbers
   (b) $f(x)=x^2;\quad g(x)=2x,\ 0\leq x$
   Answer: $(f+g)(x)=x^2+2x,\ 0\leq x$
   (c) $f(x)=1-x^2,\ \frac{1}{2}\leq x\leq 2;\quad g(x)=\frac{1}{2}$
   (d) $f(x)=\frac{1}{|x|},\ 2\leq x\leq 4;\quad g(x)=x^3,\ 1\leq x\leq 3$
   (e) $f(x)=x,\ -1\leq x\leq 1;\quad g(x)=2-x,\ 0\leq x\leq 10$
   (f) $f(x)=\sqrt{1-x^2};\quad g(x)=|x|,\ 0\leq x$
   Answer: $(f+f)(x)=|x|+\sqrt{1-x^2},\ 0\leq x\leq 1$
   (g) $f(x)=\sqrt{4-x^2};\quad g(x)=\sqrt{x^2-1}$
   (h) $f(x)=\sqrt{1+x};\quad g(x)=\sqrt{9-x^2}$
   (i) $f(x)=\begin{cases}1-x, & 0\leq x\leq 1\\ 4x, & 1<x\leq 2;\end{cases}\quad g(x)=\begin{cases}x^2, & 0\leq x\leq 1\\ x^3, & 1<x\leq 2\end{cases}$

Answer: $(f+g)(x) = \begin{cases} 1 - x + x^2, & 0 \leqslant x \leqslant 1 \\ 4x + x^3, & 1 < x \leqslant 2 \end{cases}$

(j) $f(x) = \begin{cases} x, & -2 \leqslant x < 0 \\ x+1, & 0 \leqslant x \leqslant 2; \end{cases}$ $g(x) = \begin{cases} 1, & -2 \leqslant x < 0 \\ x, & 0 \leqslant x \leqslant 2 \end{cases}$

(k) $f(x) = \begin{cases} x, & -1 \leqslant x \leqslant 0 \\ x+1, & 0 < x \leqslant 2; \end{cases}$ $g(x) = \begin{cases} 1, & -1 \leqslant x \leqslant 1 \\ x, & 1 < x \leqslant 2 \end{cases}$

(l) $f(x) = \begin{cases} |x|, & 0 \leqslant x < 2 \\ \sqrt{1+x^2}, & 2 \leqslant x \leqslant 4; \end{cases}$ $g(x) = \begin{cases} x^2, & 0 \leqslant x \leqslant 1 \\ x^3, & 1 < x \leqslant 3 \end{cases}$

(m) $f(x) = \begin{cases} x, & 1 \leqslant x \leqslant 2 \\ x^2, & 2 < x \leqslant 3; \end{cases}$ $g(x) = \begin{cases} 1, & 1 \leqslant x < 2 \\ 1/x, & 2 \leqslant x \leqslant 3 \end{cases}$

(n) $f(x) = \begin{cases} x, & 0 \leqslant x < 1 \\ x^2, & 1 \leqslant x < 2 \\ 1/x, & 2 \leqslant x \leqslant 3; \end{cases}$ $g(x) = \begin{cases} \sqrt{1-x^2}, & 0 \leqslant x \leqslant 1 \\ x, & 1 < x \leqslant 4 \end{cases}$

2. Find g in each of the following cases.
   (a) $f(x) = x;$ $(f+g)(x) = 1 + 2x$
   (b) $f(x) = x;$ $(f+g)(x) = 1$
   (c) $f(x) = |x|;$ $(f+g)(x) = 1 + x$
   (d) $f(x) = \begin{cases} 1 - x, & 0 \leqslant x \leqslant 1 \\ 4x, & 1 < x \leqslant 2; \end{cases}$ $(f+g)(x) = \begin{cases} -x, & 0 \leqslant x \leqslant 1 \\ x^2, & 1 < x \leqslant 2 \end{cases}$
   (e) $f(x) = \begin{cases} x, & -1 \leqslant x \leqslant 0 \\ x+1, & 0 < x \leqslant 2; \end{cases}$ $(f+g)(x) = \begin{cases} 0, & -1 \leqslant x \leqslant 1 \\ 1, & 1 < x \leqslant 2 \end{cases}$
   (f) $f(x) = x;$ $(f-g)(x) = x^2$
   Answer: $g(x) = x - x^2$
   (g) $f(x) = x;$ $(f-g)(x) = 1$
   (h) $f(x) = |x|;$ $(f-g)(x) = x + x^2$
   (i) $f(x) = x;$ $(fg)(x) = x^3$
   Answer: $g(x) = x^2$
   (j) $f(x) = x^2;$ $(fg)(x) = 1/x$
   (k) $f(x) = \sqrt{1-x^2};$ $(fg)(x) = |x|$
   (l) $f(x) = \sqrt{1-x^2};$ $(f/g)(x) = x$
   Answer: $g(x) = \dfrac{\sqrt{1-x^2}}{x}$
   (m) $f(x) = x;$ $(f/g)(x) = 1 - x$
   (n) $f(x) = |x|;$ $(f/g)(x) = |1 - x|$
   (o) $f(x) = \begin{cases} x, & 0 \leqslant x \leqslant 1 \\ \sqrt{1+x}, & 1 < x \leqslant 2; \end{cases}$ $(fg)(x) = \begin{cases} x^2 - x^3, & 0 \leqslant x \leqslant 1 \\ x, & 1 < x \leqslant 2 \end{cases}$
   (p) $f(x) = \begin{cases} x, & 0 \leqslant x < 1 \\ x^2, & 1 \leqslant x < 2 \\ 1/x, & 2 \leqslant x \leqslant 4; \end{cases}$ $(f/g)(x) = \begin{cases} 1, & 0 \leqslant x \leqslant 1 \\ 1/x, & 1 < x \leqslant 2 \\ 1+x, & 2 < x \leqslant 3 \\ x^2 - 1, & 3 < x \leqslant 4 \end{cases}$
3. If $g(x) = x$ and $h(x) = |x|$, write equations for and draw the graphs of $(g+h)$, $(g-h)$, (gh), and (g/h).
4. If $\phi(x) = x + \sqrt{1-x^2}$, write ϕ in the form $(f+g+h)$.
5. One way of making a sketch of the graph of $(f+g)$ is to start with the

graphs of f and g and to remember that $(f+g)(x)=f(x)+g(x)$. For any point x on the x-axis, $f(x)$ can be read off from the graph of f and $g(x)$ from the graph of g. The sum of these is then $(f+g)(x)$, and the point $(x,(f+g)(x))$ is on the graph of $(f+g)$. As many such points as necessary can be plotted and joined by a curve which is then an approximation to the graph of $(f+g)$. The same method can be used to sketch the graphs of $(f-g)$, (fg), and (f/g). Use this method to sketch the graphs of

(a) $(f+g)$ if $f(x)=x, \quad g(x)=1$
(b) $(f+g)$ if $f(x)=x, \quad g(x)=[x]$
(c) $(f-g)$ if $f(x)=x, \quad g(x)=2|x|$
(d) (fg) if $f(x)=\sqrt{1-x^2}, \quad g(x)=x$
(e) (f/g) if $f(x)=x, \quad g(x)=1+x$.

6. Find $f-g$ and f/g. Determine whether $(f-g)+g=f$ and $(f/g)g=f$ for each of

(a) $f(x)=x+1/x, \quad g(x)=1/(x-1)$
(b) $f(x)=x^2-4, \quad g(x)=x-2$.

Suppose that the domain of a function f contains a number y which is in the range of a function g. This means that $y=g(x)$ for some number x in dom g, and also that $f(g(x))$ is uniquely determined by the number x. This observation justifies the following definition of an operation, *composition,* in the class of functions.

DEFINITION 2.3 If range $g \cap \operatorname{dom} f \neq \varnothing$, then

$$f\circ g\colon (f\circ g)(x)=f(g(x)), \quad \operatorname{dom} f\circ g=\{x \mid g(x)\in \operatorname{dom} f\}.$$

The composition operation must be preceded by a sorter that selects out of the domain of g those numbers whose images, the $g(x)$'s, are in the domain of f. The entire mechanism, including the sorter, could be pictured as shown in Figure 2.25.

EXAMPLE

$$f\colon f(x)=\frac{1}{x}, \quad x\neq 0$$
$$g\colon g(x)=x-2, \quad 0\leqslant x\leqslant 2$$
$$f\circ g\colon (f\circ g)(x)=f(g(x))$$
$$=f(x-2)=\frac{1}{x-2}, \quad 0\leqslant x<2$$

Notice that although 2 is in dom g, $g(2)=0$ is not in dom f, and so 2 is sorted out of dom $f\circ g$. It seems interesting that 0 is not in dom f, but in dom $f\circ g$.

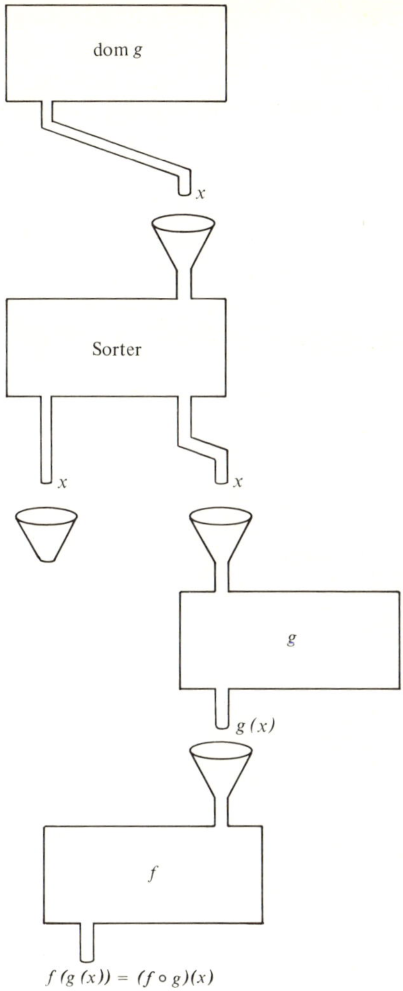

Figure 2.25

In order to perform the composition operation you may have to invent your own picture of the process, wherein you should follow through some sample computations of $(f \circ g)(x)$, such as

$$(f \circ g)(0) = f(g(0)) = f(-2) = \frac{1}{-2} = -\frac{1}{2}$$

$$(f \circ g)(1) = f(g(1)) = f(-1) = \frac{1}{-1} = -1$$

$$(f \circ g)\left(\frac{3}{2}\right) = f\left(g\left(\frac{3}{2}\right)\right) = f\left(-\frac{1}{2}\right) = \frac{1}{-1/2} = -2.$$

As we remarked earlier about sums and products of functions, trouble with domains sometimes can be avoided in making up compositions by replacing the function g in $f \circ g$ by a new function whose domain coincides with dom $f \circ g$. In this example, we might replace g by

$$g' \colon g'(x) = x - 2, \quad 0 \leqslant x < 2.$$

Then dom $f \circ g'$ = dom g'.

EXAMPLE

$$\begin{aligned} f(x) &= \sqrt{x}, & 0 \leqslant x \leqslant 4 \\ g(x) &= x(x-2), & -1 \leqslant x \leqslant 3 \\ (f \circ g)(x) &= \sqrt{x(x-2)}, & -1 \leqslant x \leqslant 0, \quad 2 \leqslant x \leqslant 3. \end{aligned}$$

See Figure 2.26 (a), (b), and (c).

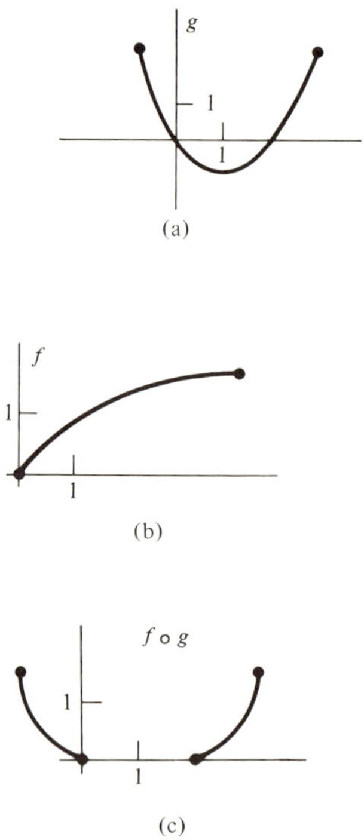

Figure 2.26

2.7 PROBLEMS (*continued*)

7. Describe $f \circ g$ if
   (a) $f(x) = x^2 - 3x, \quad g(x) = 1 - x.$

Answer: Domain consists of all numbers.

$$(f \circ g)(x) = x^2 + x - 2$$

(b) $f(x) = 1 - x, \quad g(x) = x^2 - 3x.$
(c) $f(x) = \sqrt{x}, \quad g(x) = x^3$
Answer: $(f \circ g)(x) = x^{3/2}, \quad 0 \leqslant x$
(d) $f(x) = \sqrt{x}, \quad g(x) = x^4.$
(e) $f(x) = \dfrac{|x|}{x}, \quad g(x) = 2 + x.$
(f) $f(x) = g(x) = x.$
(g) $f(x) = g(x) = x^2.$
(h) $f(x) = x^2, \quad g(x) = 1.$
(i) $f(x) = 1, \quad g(x) = x^2.$
(j) $f(x) = x, \quad 0 \leqslant x \leqslant 1; \qquad g(x) = 2x, \quad 0 \leqslant x \leqslant 1.$
Answer: $(f \circ g)(x) = 2x, \quad 0 \leqslant x \leqslant \frac{1}{2}$
(k) $f(x) = 2x, \quad 0 \leqslant x \leqslant 1; \qquad g(x) = x, \quad 0 \leqslant x \leqslant 1.$
(l) $f(x) = 2x, \quad 0 \leqslant x \leqslant 1; \qquad g(x) = 2x, \quad 0 \leqslant x \leqslant 1.$
(m) $f(x) = \begin{cases} -x, & x < 0 \\ x + \frac{1}{2}, & 0 < x; \end{cases} \qquad g(x) = \begin{cases} x^2, & x < 0 \\ -1, & 0 \leqslant x. \end{cases}$

8. Describe and draw the graph of $(f \circ g)$ if
(a) $f(x) = \sqrt{x}, \quad g(x) = 1 - x^2.$
Answer: $(f \circ g)(x) = \sqrt{1 - x^2}$
(b) $f(x) = 1 - x^2, \quad g(x) = \sqrt{x}.$
(c) $f(x) = 2 + x, \quad g(x) = 4 - x.$
(d) $f(x) = 2 + x, \quad g(x) = 2 - x.$

9. Describe $f(x)$ and a domain for g, if
(a) $g(x) = 1 - x^2, \quad (f \circ g)(x) = \sqrt{1 - x^2}.$
Answer: $f(x) = \sqrt{x}, \quad x \geqslant 0, \quad \text{dom } g = [-1,1].$
(b) $g(x) = x^2, \quad (f \circ g)(x) = \sqrt{1 - x^2}.$
(c) $g(x) = x, \quad (f \circ g)(x) = \sqrt{1 - x^2}.$
(d) $g(x) = 1 - x, \quad (f \circ g)(x) = x.$
(e) $g(x) = 2x + 3, \quad (f \circ g)(x) = 4x^2 + 12x + 9.$

10. Describe g if
(a) $f(x) = x^2, \quad (f \circ g)(x) = x^2 + 2x + 1.$
Answer: $g(x) = x + 1$
(b) $f(x) = x + 1, \quad (f \circ g)(x) = x^2.$
(c) $f(x) = 1 - x, \quad (f \circ g)(x) = x.$
(d) $f(x) = 1 + x + x^2, \quad (f \circ g)(x) = x^2 - 3x + 3.$

11. What is the domain of $(f \circ g)$ if
(a) $f(x) = \sqrt{x}, \quad g(x) = 1 + x.$
Answer: $-1 \leqslant x$

(b) $f(x) = \sqrt{x}, \quad g(x) = 1 - x^2$.
(c) $f(x) = \sqrt{x}, \quad g(x) = 1 + x^2$.
(d) $f(x) = \sqrt{x}, \quad g(x) = -x^2$.

Chapter 3 LIMIT OF A FUNCTION AND CONTINUITY

3.1 LIMIT OF A FUNCTION

There are many problems which lead to the notion of a limit of a function. When we approximate some quantity and realize that with more work our approximation can be improved, we are entering on a theory of limits, however little. Suppose that a circle is drawn on a piece of graph paper, and that we desire its area. An approximation to the area can be found by counting the squares inside the circle, and furthermore, it would appear that better approximations would be obtained if graph paper with smaller squares were used.

Often we are given a formula which seems to avoid the approximating process, such as $A = \pi r^2$ for the area of a circle with radius r. But the number π being irrational cannot be exactly represented by a decimal or by any collection of fractions, so the search for a numerical area brings us back to an approximating process. A theory of functions provides a fruitful approach to an analysis of approximation, and it is the task of this book to begin such a theory.

Returning for a moment to π, we know that the function

$$f: f(x) = 4 \arctan x, \quad x \text{ real}$$

takes the value π at $x = 1$; that is, $f(1) = \pi$. So a knowledge of the behavior of the function arctan near $x = 1$ in its domain might give us an idea of how big π is, even though we cannot write out π in figures. In Section 2.6 we examined the values of a function near a number in its domain when we studied breaks in certain graphs. We will now collect a few more examples and proceed to develop this technique.

Consider the four functions in Figure 3.1, none of which have zero in the domain. The functions h and j appear to have one property in particular near zero that neither f nor g have. For the function h there is a number A ($A = 1$) such that for each positive number ϵ, the interval $(A - \epsilon, A + \epsilon)$ contains all values, $h(x)$, of h for nonzero domain numbers x in some interval $(-\delta,\delta)$.

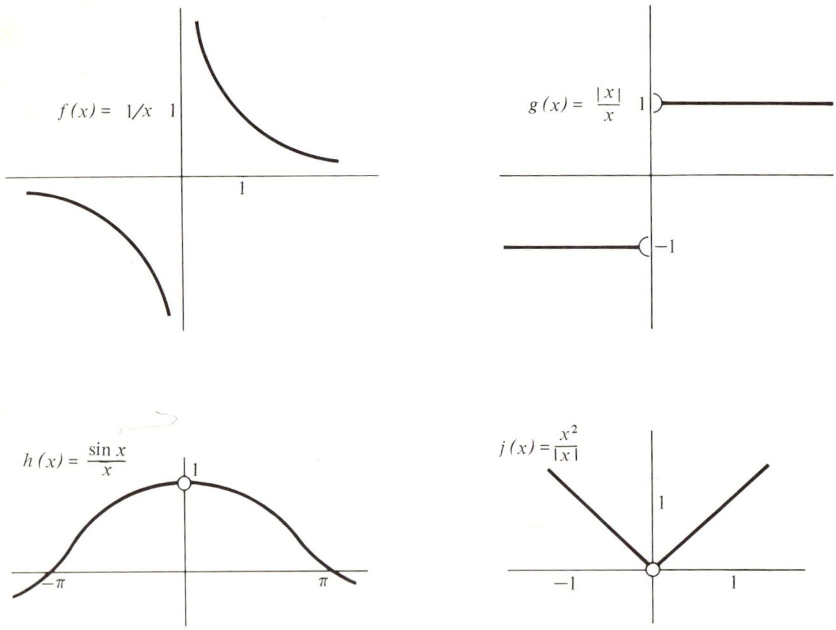

Figure 3.1

Notice that such an interval $(-\delta,\delta)$ can be found for each $\epsilon > 0$, whether ϵ is small or large. Notice also that 0 is not in the domain of h, though if it were this discussion would be unaffected. For the function j do you see that, at zero, $A = 0$?

Let us prove algebraically that both h and j have this property at 0; thus we prove for the functions themselves what their graphs strongly suggest ought to be true about them. See Figure 3.2.

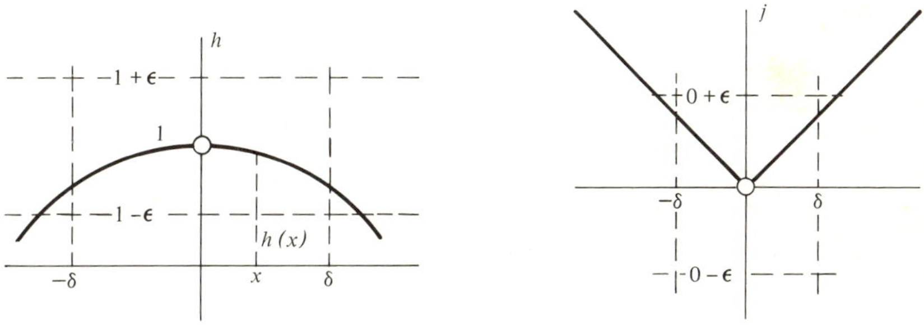

Figure 3.2

For h we need to know that for any $\epsilon > 0$ there is a number δ_ϵ such that

$$(1 - \epsilon)x < \sin x < x, \quad 0 < x < \delta_\epsilon.$$ [1]

Take $\delta = \delta_\epsilon$. Then if $x \in (-\delta,\delta)$ and $x > 0$, it follows that

$$1 - \epsilon < \frac{\sin x}{x} < 1 < 1 + \epsilon.$$

If $x \in (-\delta,\delta)$ and $x < 0$, then $\sin x/x = \sin(-x)/-x$, where $-x > 0$, and so, generally,

$$1 - \epsilon < h(x) < 1 + \epsilon, \quad -\delta < x < \delta \quad \text{with } x \neq 0,$$

which we set out to verify.

For j it is easier. Given any $\epsilon > 0$, take $\delta = \epsilon$, and then if $-\delta < x < \delta$ with $x \neq 0$,

$$|j(x) - 0| = \left|\frac{x^2}{|x|}\right| = |x| < \delta = \epsilon,$$

in which we expressed the vertical interval in absolute value notation to abbreviate the writing.

Apparently for neither of the functions f or g can such a number A be found. We postpone a verification of this impossibility, and turn to a systematic study applicable to a wide class of functions.

Consider the domain, dom f, of some function f, and a number a.

[1] An adequate definition of the sine function enables us to prove these facts about it. If the definition you know for sine does not, then accept this and read on. You will see an adequate definition of sine later on in this book.

DEFINITION 3.1 The number a is an *accumulation point* of dom f in case any open interval I, containing a, contains also a number in dom f different from a; that is,

$$[I - \{a\}] \cap \text{dom } f \neq \varnothing.$$

When we studied the function

$$h: h(x) = \frac{\sin x}{x}, \quad x \neq 0,$$

we considered an open interval $(-\delta,\delta)$, and selected a number $x \in (-\delta,\delta) \cap \text{dom } h$. Since 0 is an accumulation point of dom h, such a number x exists for any $\delta > 0$.

Let f be a function and a be an accumulation point of dom f.

DEFINITION 3.2 The number A is called the *limit* of f at a provided that for each $\epsilon > 0$ there is a $\delta > 0$ so that for each $x \in \text{dom} f$ the inequalities

$$0 < |x - a| < \delta$$

imply

$$|f(x) - A| < \epsilon.$$

Let us subsume this long definition in the symbol $\lim_a f = A$. Back in the examples, then,

$$\lim_0 h = 1, \qquad \lim_0 j = 0$$

and to neither of $\lim_0 f$ nor $\lim_0 g$ have we yet assigned a number.[2]

3.1 PROBLEMS

1. Identify all the accumulation points for each of the following sets of numbers.
   (a) the interval $[0,1]$, the interval $(0,1]$, and the interval $(0,1)$
   (b) $\{1,\frac{1}{2},\frac{1}{3},\frac{1}{4}, \ldots ,1/n, \ldots ,\}$
   (c) the union of intervals $[0,1) \cup (1,2]$

[2] As we study examples of the application of this definition, you will quickly observe that, given a number ϵ, the choice of a corresponding number δ will depend not only on ϵ but on the accumulation point a and on the function f. We could remind ourselves of this dependence by writing $\delta(\epsilon,a,f)$ for δ in each application of the definition. If this helps you understand the definition, use this symbol in your writing until the definition is ineradicably implanted in your memory.

(d) S contains $0, \frac{1}{4}, \frac{2}{4}, \frac{3}{4}$, and it contains $(2p+1)/2^{n+1}$ if it contains $p/2^n$.
(e) $\{1,2,3, \ldots, n, \ldots,\}$
(f) $\{1/10^n \mid n = 0,1,2, \ldots, 10^{100}\}$

2. Verify by Definition 3.2 that $\lim_2 f = \lim_2 g = \lim_2 h$ when

$$\begin{aligned} f(x) &= 3x, \quad 0 \leqslant x \leqslant 4 \\ g(x) &= 3x, \quad 0 \leqslant x < 2, \quad 2 < x \leqslant 4 \\ h(x) &= 3x, \quad 0 \leqslant x < 2, \quad 2 < x \leqslant 4 \\ h(2) &= 7. \end{aligned}$$

3. Show, for the functions of Problem 2, that $\lim_2 f = f(2)$, $\lim_2 h \neq h(2)$, $\lim_2 g$ exists but $g(2)$ does not.
4. If f is the function given by

$$f(x) = x - 4, \quad \operatorname{dom} f = \{1, \frac{1}{2}, \frac{1}{3}, \ldots, \frac{1}{n}, \ldots\},$$

prove that $f(0)$ does not exist and that $\lim_0 f = -4$. Why, at this stage of our work, is $\lim_a f$, $a \neq 0$, undefined?

5. Let f be the function given by

$$\begin{aligned} f(x) &= 5x \quad, \quad 0 \leqslant x < 2 \\ &= 8 - x, \quad 3 \leqslant x \leqslant 5. \end{aligned}$$

Prove that $\lim_3 f = f(3)$, $\lim_2 f = 10$, $f(2)$ does not exist.

6. Prove that if $f(x) = 5 - 2x$, then $\lim_4 f = -3$.
7. Prove that if $f(x) = |x|$, then $\lim_1 f = 1$, $\lim_{-1} f = 1$.
8. It is true that $\lim_2 f = 4$ if $f(x) = x^2$. Find a number $\delta > 0$ such that when $0 < |x - 2| < \delta$, then $|f(x) - 4| < \frac{1}{10}$.
9. Show that if f is given by $f(x) = \sqrt{x}$, $0 \leqslant x$, then for any number $a \geqslant 0$, $\lim_a f = \sqrt{a}$.

Hint: Choose $\delta = \epsilon\sqrt{a}$ and observe that

$$\sqrt{x} - \sqrt{a} = \frac{(x - a)}{\sqrt{x} + \sqrt{a}}.$$

10. Let a be an accumulation point of dom f. Define

$$\begin{aligned} f^{>}: f^{>}(x) &= f(x), \quad x > a \\ f^{<}: f^{<}(x) &= f(x), \quad x < a. \end{aligned}$$

If $\lim_a f^{>}$ exists, it is called the *right-hand* limit of f at a, and is denoted by $\lim_{a^+} f$. A similar definition of $\lim_{a^-} f$ can be made. Find right- and left-hand limits for

(a) $g(x) = \dfrac{|x|}{x}$ at 0;
(b) h, of Problem 2, at 2;
(c) f, of Problem 4, at 0 (right-hand only).

11. Prove that if $\lim_{a^+} f = \lim_{a^-} f$, then $\lim_{a} f$ exists.

This kind of procedure, that of determining a limit number at an accumulation point of the domain of a function, can never yield two answers. This is proved by the following theorem.

THEOREM 3.1 Let f be a function, and suppose that $\lim_a f = A$ and $\lim_a f = B$. Then $A = B$.

Proof: Assume that $A \neq B$; that is, $|B - A| > 0$. Let $\epsilon = |B - A|$; then $\epsilon/2 > 0$, and there is a $\delta_A > 0$ so that the inequalities

$$0 < |x - a| < \delta_A, \quad x \in \text{dom}\, f$$

imply

$$|f(x) - A| < \frac{\epsilon}{2}.$$

Also there is a number $\delta_B > 0$ so that the inequalities

$$0 < |x - a| < \delta_B, \quad x \in \text{dom}\, f$$

imply

$$|f(x) - B| < \frac{\epsilon}{2}.$$

Take $\delta = \min(\delta_A, \delta_B)$, the lesser of these two, and some x in dom f for which $0 < |x - a| < \delta$. Then

$$\begin{aligned} |B - A| &= |(f(x) - A) - (f(x) - B)| \\ &\leqslant |f(x) - A| + |f(x) - B| < \epsilon = |B - A|, \end{aligned}$$

which contradicts Axiom 01. Thus, $A = B$. ■

This theorem asserts that if a function has a limit at a number a, it can have only one limit. The graph of the example g Problem 10, Section 3.1, at first sight, suggests that g might have two limits at zero, 1, and -1, but we have now disposed of that possibility.

EXERCISE Show that $\lim_0 g = 1$ when g is redefined to be

$$g\colon g(x) = \frac{|x|}{x}, \quad x > 0.$$

THEOREM 3.2 Let f and g be functions, and let a be an accumulation point of dom $f \cap$ dom g. Suppose that this intersection is an interval except for at most a finite number of points deleted. If

$$f(x) = g(x), \quad x \neq a, \quad x \in \operatorname{dom} f \cap \operatorname{dom} g,$$

and if $\lim_a f = A$, then $\lim_a g = A$ also.

Proof: Choose $\epsilon > 0$; then take $\delta > 0$ so that the inequalities

$$0 < |x - a| < \delta$$

imply

$$|f(x) - A| < \epsilon.$$

Since a is an accumulation point of dom g also, and

$$|g(x) - A| = |f(x) - A| < \epsilon,$$

it follows[3] that $\lim_a g = A$. ■

We decided that $\lim_0 j = 0$ (page 66). Therefore, by Theorem 3.2 such functions as

$$\begin{aligned} k\colon k(x) &= |x|, \quad x \text{ real} \\ l\colon l(x) &= \frac{x^2}{|x|}, \quad x \neq 0 \\ &= 5\ , \quad x = 0 \end{aligned}$$

also have the limit 0 at zero.

[3] Without the condition on dom $f \cap$ dom g, Theorem 3.2 would be false. Consider the example

$$f\colon f(x) = x + 1, \quad -1 \leqslant x \leqslant 1, \quad x \neq \frac{1}{n}, \quad n = 1,2,3, \ldots,$$

$$\begin{aligned} g\colon g(x) &= 2, \quad x = \frac{1}{n}, \quad n = 1,2,3, \ldots, \\ &= f(x) \quad \text{otherwise}, \end{aligned}$$

in which $\lim_0 f = 1$, but $\lim_0 g$ does not exist.

EXERCISE Do Problem 2 of page 67 using Theorem 3.2.

The next theorem contains a procedure for verifying that a number B is not the limit of a given function at some accumulation point of its domain. With it we can prove that neither f nor g has a limit at 0.

THEOREM 3.3 Let the number a be an accumulation point of the domain of a function f. The number B is not the limit of f at a in case there is some number $\epsilon > 0$ such that for any number $\delta > 0$, there is some number x in dom f for which $0 < |x - a| < \delta$ and $|f(x) - B| \geqslant \epsilon$.

Proof: We simply negate[4] the condition that $B = \lim_a f$, given in Definition 3.2. ■

Let us apply Theorem 3.3 to the function f, described on page 64, $f(x) = 1/x$, at 0. Suppose that $B > 0$. Take $\epsilon = B$ and any $\delta > 0$. There is a number, for example, $x = -\delta/2$, in dom f, for which

$$0 < |x - 0| = \left|-\frac{\delta}{2}\right| = \frac{\delta}{2} < \delta$$

and

$$|f(x) - B| = \left|f\left(-\frac{\delta}{2}\right) - B\right| = \left|-\frac{2}{\delta} - B\right| = B + \frac{2}{\delta} > B = \epsilon.$$

Therefore, $B \neq \lim_0 f$.

Suppose that $B < 0$; take $\epsilon = |B|$ and any $\delta > 0$. There is a number, for example, $x = \delta/2$, in dom f for which

$$0 < |x - 0| = \left|\frac{\delta}{2}\right| = \frac{\delta}{2} < \delta,$$

and

[4] To negate the proposition which constitutes Definition 3.2 is not difficult; it is just tedious. In Definition 3.2 we find the proposition:

(α): each $\epsilon > 0$ implies that there exists a $\delta > 0$ such that p is true, where p is a proposition. The negation of (α) is:
(not-α): There exists an $\epsilon > 0$ such that each $\delta > 0$ implies not-p is true; that is, p is false. This should be the assertion of Theorem 3.3.

To see that it is, observe that the proposition p is:

(p): each $x \in$ dom f, with $0 < |x - a| < \delta$ implies $|f(x) - B| < \epsilon$, and its negation is:
(not-p): There exists an $x \in$ dom f, with $0 < |x - a| < \delta$ such that $|f(x) - B| \nless \epsilon$.

Now reread the proposition not-α, substituting not-p, and you will get Theorem 3.3.

$$|f(x) - B| = \left|f\left(\frac{\delta}{2}\right) - B\right| = \left|\frac{2}{\delta} - B\right| = \left|\frac{2}{\delta} + |B|\right| = |B| + \frac{2}{\delta} > |B| = \epsilon.$$

Therefore, $B \neq \lim_0 f$.

If $B = 0$, take $\epsilon = 1$ and any $\delta > 0$. There is a number x, for example, $x = \min(\delta/2, 1/2)$, in dom f for which

$$0 < |x - 0| \leqslant \frac{\delta}{2} < \delta,$$

and

$$\begin{aligned} |f(x) - B| = |f(x)| &\geqslant 2 > 1 = \epsilon, \quad \left(\text{if } x = \frac{1}{2}\right) \\ &= \frac{2}{\delta} \geqslant 2 > 1 = \epsilon \quad, \quad \left(\text{if } x = \frac{\delta}{2}, \quad \text{so } \delta < 1\right). \end{aligned}$$

Therefore, $B \neq \lim_0 f$.

By Axiom O1 there are no other possibilities to examine for the number B, and the function f has no limit at 0.

Let us apply Theorem 3.3 to the function g: $g(x) = |x|/x$ described on page 64, at 0. Suppose that $B = 1$; take $\epsilon = 1$ and any $\delta > 0$. There is a number x, for example, $x = -\delta/2$ in dom g for which

$$0 < |x - 0| = \left|-\frac{\delta}{2}\right| = \frac{\delta}{2} < \delta,$$

and

$$|g(x) - B| = |-1 - 1| = 2 > 1 = \epsilon.$$

Therefore, $B \neq \lim_0 g$.

Suppose that $B \neq 1$; take $\epsilon = \frac{1}{2}|1 - B|$ and any $\delta > 0$. There is a number x, for example, $x = \delta/2$, in dom g for which

$$0 < |x - 0| = \frac{\delta}{2} < \delta$$

and

$$|g(x) - B| = |1 - B| > \frac{1}{2}|1 - B| = \epsilon.$$

Therefore, $B \neq \lim_0 g$.

Thus the function g has no limit at 0. Notice that if we modify g to be g_1,

$$\begin{aligned} g_1\colon g_1(x) &= \frac{|x|}{x}, \quad x \neq 0 \\ &= k \quad, \quad x = 0 \quad (\text{for any number } k), \end{aligned}$$

then the preceding discussion would be unaffected. Also, if $\lim_0 g_1$ exists, then so does $\lim_0 g$ by Theorem 3.2, and on both counts g_1 has no limit at 0.

We will need, later on, (see Definition 3.5) to have an agreement about the existence of the limit of a function at a number in its domain which is not an accumulation point of the domain. Let $a \in \operatorname{dom} f$ and I be an open interval containing a but no other number in $\operatorname{dom} f$. Call a an *isolated point* of $\operatorname{dom} f$. See Figure 3.3.

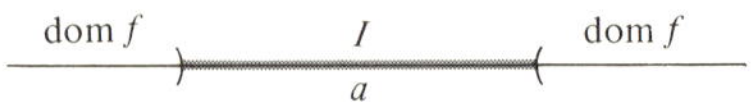

Figure 3.3

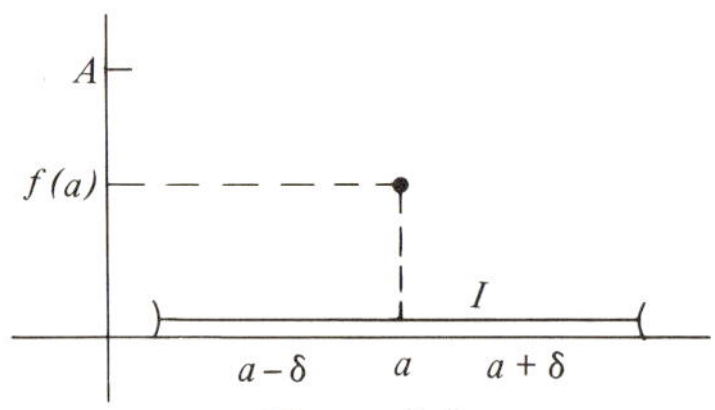

Figure 3.4

Let us test the truth of $\lim_a f = A$, for some A, by an appeal to Theorem 3.3. For whatever number ϵ we start with, there will be a δ (as indicated in Figure 3.4), so there does not exist an $x \in \operatorname{dom} f$ that close to a. Thus, by Theorem 3.3 the negation of Definition 3.2 is false for the number A. We conclude that Definition 3.2 is satisfied for A. Fortunately, A was unspecified, so we take the liberty of defining $\lim_a f = f(a)$ at an isolated point of the domain. Without this definition of $\lim_a f$ any two numbers A and B could be limits of f at a, in conflict with Theorem 3.1.

3.1 PROBLEMS (*continued*)

12. Consider

$$f\colon f(x) = 1, \quad 0 < x$$
$$\qquad\quad = 0, \quad x \leqslant 0.$$

(a) Prove that $0 \neq \lim_0 f$.

(b) Prove that $1 \neq \lim_0 f$.

(c) Prove that $\lim_0 f \neq k$ for any k.

13. Let the function f be defined by the equation

$$f(x) = \begin{cases} 1, & 0 \leqslant x < 1 \\ 2, & 1 \leqslant x \leqslant 2. \end{cases}$$

Show that $\lim_1 f$ does not exist but that the right- and left-hand limits do exist.

14. Let the function f be defined by the equation

$$f(x) = \begin{cases} 3x, & 0 \leqslant x \leqslant 1 \\ 5 - 3x, & 1 < x \leqslant 2. \end{cases}$$

Show that $\lim_{1} f$ does not exist but that the right- and left-hand limits do exist.

15. If $f(x) = 1/x$, show that neither $\lim_{0} f$ nor the right- and left-hand limits exist.

16. Returning to the function of page 64, define a new function ϕ which has the same rule as g but whose domain consists of that part of the domain of g to the right of 0; that is, $\phi(x) = 1, 0 < x$. Show that $\lim_{0} \phi = 1$.

17. Consider the function f of Problem 15 on page 53. Prove that
    (a) if a is irrational, then $\lim_{a} f = 0$;
    (b) if a is rational, then $\lim_{a} f$ does not exist.

3.2 VELOCITY

To calculate the velocity of a particle as it moves on a line is a problem which typifies the application of functions to physical phenomena. Near the surface of the earth it is observed that if a body falls freely from rest s feet in t seconds, then

$$s = 16t^2.$$

We construct the function f

$$f: f(t) = 16t^2, \quad 0 \leqslant t \leqslant 4$$

in which the domain is a set of times measured by a clock and the range is a set of distances measured along the path of fall. It is verified by observation that the rule corresponds closely to the observed event.

Given this function, one can ask for the velocity of the particle at t_0 seconds after the start. If the usual definition of velocity, that is, the ratio of the distance traveled to the time elapsed, is applied at $t = t_0$, then this ratio is represented by 0/0, a symbol we have agreed not to use (see Problem 7, Section 1.2).

However, the formula for the slope function for f at t_0,

$${}_f\sigma_{t_0}(h) = \frac{1}{h}\,[f(t_0 + h) - f(t_0)] = \frac{h}{h}\,[32t_0 + 16h], \quad h \neq 0,$$

represents the distance traveled to the time elapsed ratio from t_0 to $t_0 + h$, and this slope function may have a limit at 0. By Theorem 3.2,

$$\lim_{0} {}_f\sigma_{t_0} = \lim_{0} g,$$

where

$$g\colon g(h) = 32t_0 + 16h,$$

and $\lim_{0} g = 32t_0$, as we shall see in the next section. You can discover this for yourself by applying Definition 3.2.

This limit, $32t_0$, is used in place of the ratio 0/0, to measure the velocity of a falling body.

We proceed next with a systematic introduction to a theory of the limit of a function. This will be based on Definition 3.2, but once we do this we can use limits to describe certain properties of functions without direct appeal to that definition.

3.3 THEOREMS ON THE LIMIT OF A FUNCTION

The first two theorems are about special functions which are both useful and simple, and the proofs provide a warmup for the somewhat more involved proofs to follow. In these theorems we assume that the domain of each function contains an interval (open or closed) and, if $\lim_{a} f$ is considered, a is an accumulation point of dom f.

THEOREM 3.4 Let f be the constant function

$$f\colon f(x) = c, \quad c \text{ is a real number.}$$

Then $\lim_{a} f = c$.

Proof: If $\epsilon > 0$, take any $\delta > 0$. Then if

$$0 < |x - a| < \delta,$$

it follows that

$$|f(x) - c| = |c - c| = 0 < \epsilon. \blacksquare$$

THEOREM 3.5 Let I be the identity function

$$I\colon I(x) = x.$$

Then

$$\lim_{a} I = a.$$

Proof: If $\epsilon > 0$, take $\delta = \epsilon$. Then if $0 < |x - a| < \delta$, it follows that $|I(x) - a| = |x - a| < \delta = \epsilon$. $\blacksquare$

Polynomial functions are very useful, and any polynomial function can be constructed from constant functions and the identity function by means of arithmetic operations. Furthermore, rational functions, quotients of polynomial functions as described on page 82 can also be built up by sequences of arithmetic operations on constant functions and the identity function.

EXAMPLE If the functions K, L, and I are defined by $K(x) = 5$, $L(x) = -2$, and $I(x) = x$, then a polynomial function P is defined by $P = KI^2 + L$, $P(x) = K(x)I^2(x) + L(x) = 5x^2 - 2$. We can assume that the domains of all of these functions are the set of real numbers.

EXAMPLE With the functions given above, and

$$Q\colon Q(x) = x^2 - 1,$$

the function defined by $R(x) = (5x^2 - 2)/(x^2 - 1)$ is $R = P/Q$. In this case, since $Q(1) = Q(-1) = 0$, the domain of R does not contain either 1 or -1.

Now we investigate how to determine the limit of a polynomial or a rational function by observing how limits behave under arithmetic operations. Our results will, in fact, have a much broader application than this.

THEOREM 3.6 If f is a function and c is a real number, where $\lim_a f$ exists, then

$$\lim_a cf = c \lim_a f.$$

Proof: Let $\lim_a f = L$. Take $\epsilon > 0$. If $c = 0$, then cf is a constant function, and by Theorem 3.4,

$$\lim_a cf = 0 = 0L = c \lim_a f.$$

If $c \neq 0$, then $\epsilon/|c| > 0$, and by hypothesis there is a $\delta > 0$ such that if $0 < |x - a| < \delta$, then

$$|cf(x) - c \lim_a f| = |c|\,|f(x) - L| < |c| \frac{\epsilon}{|c|} = \epsilon. \blacksquare$$

Admonition: In each of the proofs which depend directly on the definition of the limit of a function, you should identify

(a) the function whose limit is to be checked (cf in Theorem 3.6),

(b) the conjectured limit ($c \lim_a f$ in Theorem 3.6),

(c) and a verification that the conditions of Definition 3.2 are fulfilled (in the proof of Theorem 3.6 the statement: for any $\epsilon > 0$ there is a $\delta > 0$, such that if $0 < |x - a| < \delta$, then $|cf(x) - c \lim_a f| < \epsilon$ is justified, where the number x used in the first inequality must be chosen from dom cf). An understanding of such proofs comes in reading them, and we exhibit a few more.

THEOREM 3.7 If f and g are functions, where both $\lim_a f$ and $\lim_a g$ exist, then

$$\lim_a (f+g) = \lim_a f + \lim_a g.$$

Proof: Let $\lim_a f = F$ and $\lim_a g = G$. Take $\epsilon > 0$. Then $\epsilon/2 > 0$, and by hypothesis there is a $\delta_f > 0$ so that if $0 < |x-a| < \delta_f$, then $|f(x) - F| < \epsilon/2$; and there is a $\delta_g > 0$, so that if $0 < |x-a| < \delta_g$, then $|g(x) - G| < \epsilon/2$. Choose $\delta = \min(\delta_f, \delta_g)$. If $0 < |x-a| < \delta$, both of the inequalities $0 < |x-a| < \delta_f$ and $0 < |x-a| < \delta_g$ are true, and therefore

$$|(f+g)(x) - (\lim_a f + \lim_a g)| = |(f(x) - F) + (g(x) - G)|$$
$$\leqslant |f(x) - F| + |g(x) - G| < \frac{\epsilon}{2} + \frac{\epsilon}{2} = \epsilon.$$[5] ■

Corollary to Theorem 3.7: $\lim_a (f-g) = \lim_a f - \lim_a g.$

Proof: By definition, $f - g = f + (-1)g$. Therefore, Theorems 3.6 and 3.7 imply

$$\lim_a (f-g) = \lim_a (f + (-1)g) = \lim_a f + \lim_a (-1)g = \lim_a f - \lim_a g.$$ ■

DEFINITION 3.3[6] A function f is *bounded* on an interval I if there is a positive number B such that for each $x \in I \cap \operatorname{dom} f$,

$$|f(x)| \leqslant B.$$

Lemma 1 to Theorem 3.8: If $\lim_a f$ exists, then there is an interval I, containing a, on which f is bounded.

Proof: Let $\lim_a f = F$. First, suppose that $F = 0$. By Definition 3.2, for $\epsilon = 1$ there is a $\delta > 0$ such that if $0 < |x-a| < \delta$, then

$$|f(x)| = |f(x) - F| < \epsilon = 1.$$

If $a \in \operatorname{dom} f$, then $|f(a)|$ exists. In Definition 3.3 take

[5] In light of the admonition just preceding Theorem 3.7 we see that the function whose limit is to be checked is $f+g$, the conjectured limit is the number $\lim_a f + \lim_a g$, and the proof contains:

Take $\epsilon > 0$ . .(I). . Choose $\delta = \min(\delta_f, \delta_g)$; then if $0 < |x-a| < \delta$, . .(II). . Therefore, $|(f+g)(x) - (\lim_a f + \lim_a g)|$ . .(III). . $< \epsilon$. The gap (I) justifies the possibility of this choice of δ, and gap (II) justifies the validity of the string of inequalities in gap (III).

[6] The letter I has also represented the identity function. Always be sure that the meaning of a symbol is clear to you from its context.

$$I = \{x \mid |x - a| < \delta\}, \quad B = \max(1, |f(a)|)$$

and the lemma is proved for this case.

Secondly, suppose that $F \neq 0$ so that $\frac{1}{2}|F| > 0$. By the existence of $\lim_a f$ there is a $\delta > 0$ so that if $0 < |x - a| < \delta$, then

$$|f(x) - F| < \tfrac{1}{2}|F|.$$

Therefore,

$$\big||f(x)| - |F|\big| < \tfrac{1}{2}|F|.$$

That is,

$$0 < \tfrac{1}{2}|F| < |f(x)| < \tfrac{3}{2}|F|.$$

Again, if $a \in \operatorname{dom} f$, we take

$$I = \{x \mid |x - a| < \delta\}, \quad B = \max\left(\tfrac{3}{2}|F|, |f(a)|\right)$$

as an interval and bound in Definition 3.3. ■

Lemma 2 to Theorem 3.8: Let $\lim_a f = F$ and let ϕ be the constant function

$$\phi\colon \phi(x) = F.$$

Then $\lim_a(\mathrm{f} - \phi) = 0$ if, and only if, $\lim_a f = F$.

Proof: Take $\epsilon > 0$; then there is a $\delta > 0$ so that if

$$0 < |x - a| < \delta,$$

it follows that

$$|f(x) - F| = |(f - \phi)(x) - 0| < \epsilon$$

in case $\lim_a(f - \phi) = 0$. This proves the "only if" part of the lemma. Looked at the other way around, take $\epsilon > 0$; then there is a $\delta > 0$ so that if $0 < |x - a| < \delta$, then it follows that

$$|(f - \phi)(x) - 0| = |f(x) - F| < \epsilon$$

in case $\lim_a f = F$. ■

Lemma 3 to Theorem 3.8: If $\lim_a f = 0$ and a function g is bounded on an interval I containing a in its interior, then $\lim_a(fg) = 0$.

Proof: By Definition 3.3 there is a number $B > 0$ so that

$$|g(x)| \leq B, \quad x \in I.$$

Take $\epsilon > 0$; then there is a $\delta_1 > 0$ so that if $0 < |x - a| < \delta_1$, $|f(x)| <$

ϵ/B. Since a is interior to I, there is a $\delta > 0$ so that $\delta \leqslant \delta_1$, and $x \in \{x|\ |x-a| < \delta\}$ implies $x \in I$. Thus, if $0 < |x-a| < \delta$, it follows that

$$|(fg)(x) - 0| = |f(x)|\ |g(x)| < \frac{\epsilon}{B} B = \epsilon. \blacksquare$$

THEOREM 3.8 If f and g are functions, where both $\lim_a f$ and $\lim_a g$ exist, then

$$\lim_a (fg) = \lim_a f \lim_a g.$$

Proof: Let $\lim_a f = F$ and $\lim_a g = G$, and let ϕ and γ be the constant functions

$$\phi\colon \phi(x) = F \qquad \gamma\colon \gamma(x) = G.$$

Observe the function identity

$$\begin{aligned} fg - \phi\gamma &= fg - f\gamma + f\gamma - \phi\gamma \\ &= f(g-\gamma) + (f-\phi)\gamma, \end{aligned}$$

in which f is bounded in some interval I_f containing a, since $\lim_a f$ exists, and $\lim_a (g-\gamma) = 0$ (see Lemmas 1 and 2). Therefore, by Lemma 3, $\lim_a f(g-\gamma) = 0$. Also $\lim_a (f-\phi) = 0$ (see Lemma 2), and γ is clearly bounded; therefore, by Lemma 3, again, $\lim_a (f-\phi)\gamma = 0$. There is a common interval, containing a, on which these functions are bounded. Hence, by Theorem 3.7

$$\lim_a (fg - \phi\gamma) = \lim_a f(g-\gamma) + \lim_a (f-\phi)\gamma = 0 + 0 = 0$$

and by Lemma 2 and Theorem 3.4,

$$\lim_a fg = \lim_a \phi\gamma = FG = \lim_a f \lim_a g. \blacksquare$$

Notice that most of the detailed work was done in the lemmas, and that the proof of Theorem 3.8 consisted essentially of stating the identity which broke the problem down into pieces.

THEOREM 3.9 Let g be a function and let $\lim_a g$ exist and be not 0. Then

$$\lim_a \frac{1}{g} = \frac{1}{\lim_a g}.$$

Proof: During the proof of Lemma 1 to Theorem 3.8, we observed incidentally that a function such as g is bounded away from 0 in absolute value. That is, if $\lim_a g = G \neq 0$, then for some $\delta > 0$,

$$|g(x)| > \tfrac{1}{2}|G|, \quad 0 < |x-a| < \delta.$$

We call this set of numbers $\{x|0 < |x-a| < \delta\}$ the *deleted* open interval at a. As before, define the constant function

$$\gamma\colon \gamma(x) = G.$$

In the deleted open interval at a, then,

$$\left|\frac{-1}{Gg(x)}\right| < \frac{2}{|G|^2}$$

so the function $-1/\gamma g$ is bounded. With the hypothesis $\lim\limits_a g = G$, Lemma 2 to Theorem 3.8 gives $\lim\limits_a (g-\gamma) = 0$. Thus, the function identity

$$\frac{1}{g} - \frac{1}{\gamma} = \frac{-1}{\gamma g}(g-\gamma),$$

with the aid of Lemma 3 to Theorem 3.8, gives the conclusion

$$\lim_a \left(\frac{1}{g} - \frac{1}{\gamma}\right) = \lim_a \frac{-1}{\gamma g}(g-\gamma) = 0,$$

and hence $\lim\limits_a (1/g) = (1/G)$ by Lemma 2 to Theorem 3.8. ■

Corollary to Theorem 3.9: Let f and g be functions where both $\lim\limits_a f$ and $\lim\limits_a g$ exist with $\lim\limits_a g$ not 0. Then

$$\lim_a \frac{f}{g} = \frac{\lim\limits_a f}{\lim\limits_a g}.$$

Proof: Applying Theorems 3.8 and 3.9 to the function identity

$$\frac{f}{g} = f\left(\frac{1}{g}\right),$$

we see immediately that

$$\lim_a \frac{f}{g} = \lim_a f\left(\frac{1}{g}\right) = \lim_a f \lim_a \frac{1}{g} = \lim_a f\left(\frac{1}{\lim\limits_a g}\right). \;■$$

3.3 PROBLEMS

1. Apply the theorems of this section to find each of the following limits, $\lim\limits_a f$.

   (a) $f(x) = 4x + 3, \quad a = 3$

(b) $f(x) = 4x^2 - 2, \quad a = 2$
(c) $f(x) = x^2 - 2x + 3, \quad a = 10$
(d) $f(x) = 2x + \sqrt{x}, \quad a = 4$
(e) $f(x) = x^2\sqrt{x} + 2\sqrt{x}, \quad a = 0$
(f) $f(x) = (x^2 + 2)(\sqrt{x} - 1), \quad a = 9$
(g) $f(x) = (x^2 + 1)/(2 - x), \quad a = 1$
(h) $f(x) = (2x^2 - x + 3)/(1 - x), \quad a = 3$
(i) $f(x) = (1 + x)(1 + x^2)(2 + \sqrt{x})/(1 - 2x)(2 - x), \quad a = 5$
(j) $f(x) = 1 + |x|, \quad a = 1$
(k) $f(x) = (1 + |x|)/x^2, \quad a = -1$
(l) $f(x) = (a\sqrt{x} + x\sqrt{a})/(a + x), \quad a > 0$

2. Find $\lim_a f$ for each of the following cases (recall Theorem 3.2).
(a) $f(x) = (2x - 2x^2)/(1 - x), \quad a = 1$
(b) $f(x) = (x^2 - x)/2x, \quad a = 0$
(c) $f(x) = (x^3 - 1)/(x - 1), \quad a = 1$
(d) $f(x) = (x^2 - 2x - 3)/(x^2 - 4x + 3), \quad a = 3$
(e) $f(x) = (x - 4)/(x^2 - 16), \quad a = 4$

3. Proceed as in Problem 2 for the following.
(a) $f(x) = (\sqrt{x} - 2)/(x - 4), \quad a = 4$
Hint: Factor the denominator.
(b) $f(x) = (1 - x)/(1 - \sqrt{x}), \quad a = 1$
Hint: Factor the numerator.
(c) $f(x) = (x^2 - 3)/(x - \sqrt{3}), \quad a = \sqrt{3}$
(d) $f(x) = (\sqrt{x} - \sqrt{2})(x^2 - 2x), \quad a = 2$

4. For each function f below, define the slope function at $x \in \operatorname{dom} f$ by the formula

$$_f\sigma_x(h) = \frac{1}{h}\,[f(x + h) - f(x)], \quad x + h \in \operatorname{dom} f.$$

Find $\lim_0 {_f\sigma_x}$ for each of the following cases.
(a) $f(x) = 2x - x^2$
Hint: $_f\sigma_x(h) = (1/h)[2h - 2xh - h^2]$, and $\lim_0 {_f\sigma_x} = 2 - 2x$, $x \in \operatorname{dom} f$. Notice $\lim_0 {_f\sigma_0} = 2$, $\lim_0 {_f\sigma_1} = 0$, and so on.
(b) $f(x) = x^2$
(c) $f(x) = x^3$
(d) $f(x) = x^4 - x^2$
(e) $f(x) = x/(1 + x)$
(f) $f(x) = \sqrt{x} - x$

5. Find the velocity of a particle at time t_0 for each of the following displacement functions g.
(a) $g(t) = 4t, \quad t_0 = 2$

(b) $g(t) = t^2 - 2t, \quad t_0 = 3$
(c) $g(t) = t/(1 + t), \quad t_0 = 5$
(d) $g(t) = \sqrt{t}, \quad t_0 = 100$

3.4 RATIONAL FUNCTIONS

We note in passing that polynomial functions have little to contribute to the study of limits. As we shall see, however, the theory of limits does contribute significantly to a study of polynomial functions.

The domain of a polynomial function P is the set of all real numbers, and its rule is[7]

$$P\colon P(x) = \sum_{j=0}^{n} c_j x^j, \quad \{c_j | j = 0, 1, \ldots, n\} \quad \text{is a set of real numbers.}$$

We are looking at a particular example of a function, and it is a handy device to use the rule for a function as a name for it, as we observed in Section 2.3, especially when the lim symbol is involved. An example will make this clear.

If P is the function

$$P\colon P(x) = 4x^3 - 7, \quad \text{all real } x,$$

we can calculate $\lim_{2} P = 25$. The essential information here is also conveyed by the symbol $\lim_{x \to 2}(4x^3 - 7) = 25$. This is a hybrid symbol; we have defined the limit of a function not of a number. The x in the subscript has its use in this hybrid, for without it, consider $\lim_{2}(lx^2 + m)$. Does this refer to the linear function

$$f\colon f(l) = x^2 l + m, \text{ for some } x \text{ and } m$$

or to the quadratic function

$$g\colon g(x) = lx^2 + m, \quad \text{for some } l \text{ and } m?$$

The subscript designates the letter in the formula that stands for a number in the domain of the implied function. The arrow reminds us that we are, in a special way, interested in values of the function near an accumulation point of its domain (2, in this case). This notation is widely used.

Using this convention for the polynomial function P, we note that if $n = 0$,

[7] See Appendix II for an explanation of the sum notation. Notice that if I denotes the identity function, the polynomial function is

$$P = \sum_{j=0}^{n} c_j I^j.$$

$$\lim_{x \to a} P(x) = \lim_{x \to a} \sum_{j=0}^{n} b_j x^j = \lim_{x \to a} b_0 = b_0 = P(a)$$

by Theorem 3.4. Suppose that for every polynomial function P with degree less than n,

$$\lim_{x \to a} P(x) = P(a).$$

Then for a polynomial function with degree n,

$$\begin{aligned} \lim_{x \to a} P(x) = \lim_{x \to a} \sum_{j=0}^{n} b_j x^j &= \lim_{x \to a} \left[\sum_{j=0}^{n-1} b_j x^j + b_n x^n \right] \\ &= \lim_{x \to a} \sum_{j=0}^{n-1} b_j x^j + \lim_{x \to a} b_n x^n, \end{aligned}$$

by Theorem 3.7, if the $\lim_{x \to a} b_n x^n$ exists. Using a selection of the theorems we have proved, and mathematical induction, you can prove that it does, and is

$$\lim_{x \to a} b_n x^n = b_n a^n.$$

Thus

$$\lim_{x \to a} P(x) = \sum_{j=0}^{n-1} b_j a^j + b_n a^n = P(a),$$

where P has degree n. There is no problem in calculating the limit of a polynomial function at a number a; simply compute $P(a)$.[8]

A function R is a rational function if there exist two polynomial functions P and Q such that

$$R(x) = \frac{P(x)}{Q(x)}$$

for all numbers x in dom R. Notice that Q must be chosen so that no number in dom R is a zero of Q.

If $a \in$ dom R, then $\lim_a Q \neq 0$ (see above), and our theorems give at once

$$\lim_a R = \lim_a \frac{P}{Q} = \frac{\lim_a P}{\lim_a Q} = \frac{P(a)}{Q(a)} = R(a).$$

On the other hand, if $a \notin$ dom R, but is an accumulation point of dom R then, of course, $R(a)$ does not exist. But $\lim_a R$ may exist as the following

[8] This illustration was written in the "arrow" notation to introduce its use. In function notation we proved that for any polynomial function P, $\lim_a P = P(a)$.

example shows. Let R be the function

$$R\colon R(x) = x^2 + x + 1, \quad 0 \leqslant x < 1, \quad 1 < x \leqslant 2,$$

which is not a polynomial function (see the domain). However, in dom R,

$$R(x) = \frac{x^3 - 1}{x - 1}, \quad x \in \text{dom } R.$$

Since $1 \notin$ dom R, $R(1)$ does not exist, but

$$\lim_{1} R = \lim_{x \to 1} \frac{x^3 - 1}{x - 1} = \lim_{x \to 1} (x^2 + x + 1) = 3$$

by Theorem 3.2. This function R has a limit when it does not have a value. We saw this happen before in the section on velocity, and we shall see in the next chapter that this phenomenon is a cornerstone of the theory of the derivative.

We conclude this section with three useful theorems, which you can prove.

THEOREM 3.10 If $\lim_a f > 0$ (<0), there is an interval I, containing a, such that for each $x \in I \cap \text{dom } f - \{a\}, f(x) > 0 (<0)$.

THEOREM 3.11 Suppose that both $\lim_a f$ and $\lim_a g$ exist. If there is an interval I, containing a, in which $f(x) \leqslant g(x)$, then $\lim_a f \leqslant \lim_a g$.

THEOREM 3.12 Suppose that $\lim_a f = \lim_a g$. If there is an interval I, containing a, in which $f(x) \leqslant h(x) \leqslant g(x)$, then $\lim_a h$ exists and $\lim_a h = \lim_a f$.

3.4 PROBLEMS

1. Write out formulas for each of the following polynomial functions, $P(x) = \sum_{j=0}^{n} c_j x^j$.
   (a) $c_j = 1/(j + 1), \quad n = 3$
   (b) $c_j = j!, \quad n = 4$
   (c) $c_j = (-1)^j j^2, \quad n = 5$
2. Find $\lim_2 P$ for each of the functions in Problem 1.
3. Prove that $\lim_0 P = c_0$ for any polynomial function.
4. (a) Find each of

$$\lim_{x \to 3} (ax^2 + b^2x), \quad \lim_{a \to 3} (ax^2 + b^2x), \quad \lim_{b \to 3} (ax^2 + b^2x).$$

(b) Describe a function f for each case in (a) so that the limit can be expressed as $\lim_3 f$.

5. Prove by mathematical induction that $\lim_{x \to a} cx^n = ca^n$, n a nonnegative integer.
6. Which of the following are rational functions?

   (a) $f(x) = x - 2, \quad x \neq -2$

   Hint: Examine $\dfrac{x^2 - 4}{x + 2}$.

   (b) $f(x) = \sqrt{x}, \quad x \geqslant 0$

   Hint: If two polynomial functions take the same value at each number in an interval, they have the same degree. Therefore $\sqrt{x}$ is not a rational function.
7. Show why every polynomial function is a rational function.
8. Prove that $\lim_{x \to 3} x/(x - 3)$ does not exist. Conclude that a rational function does not necessarily have a limit at an accumulation point of its domain.
9. Prove Theorem 3.10. Show by a counterexample that if the symbol $\{a\}$ is erased in its statement, the resulting theorem is false.
10. Prove Theorem 3.11. Construct an example for which $f(x) < g(x)$, in I, but $\lim_a f = \lim_a g$.
11. Prove Theorem 3.12. Suppose that $\lim_a f = \lim_a g$ and $f(x) \leqslant h(x) \leqslant g(x)$ for those numbers $x \in \operatorname{dom} f \cap \operatorname{dom} g \cap \operatorname{dom} h$. Under this supposition, construct a function h for which $\lim_a h$ does not exist.

3.5 CONTINUOUS FUNCTION

Let us modify slightly the illustrative functions of Section 3.1, and consider the four functions of Figure 3.5.

$$
\begin{aligned}
f\colon f(x) &= \frac{1}{x}, && x \neq 0 \\
&= 1, && x = 0 \\
g\colon g(x) &= \frac{|x|}{x}, && x \neq 0 \\
h\colon h(x) &= \frac{\sin x}{x}, && x \neq 0 \\
&= \frac{1}{2}, && x = 0 \\
j\colon j(x) &= |x|, && x \text{ real}
\end{aligned}
$$

We state the following definition.

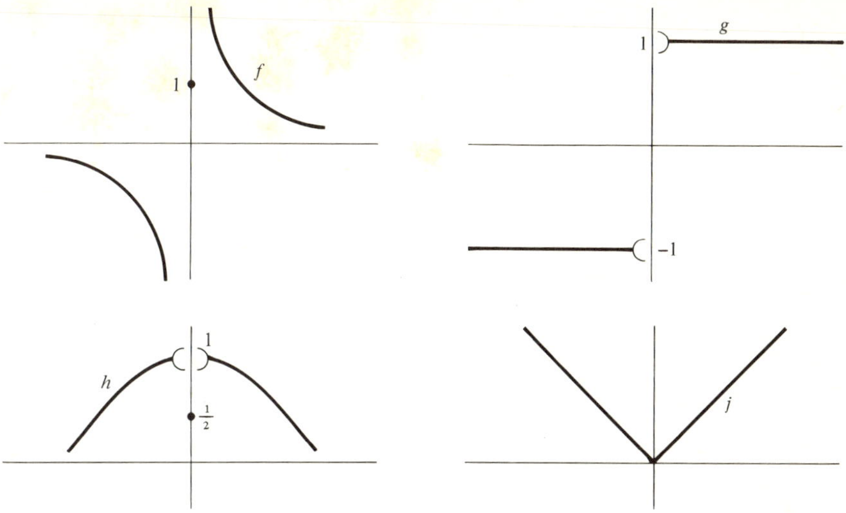

Figure 3.5

DEFINITION 3.4 A function f is *continuous at the number a* if $\lim_{a} f = f(a)$.[9]

Only one of the functions in Figure 3.5 is continuous at 0. The function f has no limit at 0. The function g has neither a limit nor a value at 0. For the function h, $\lim_{0} h = 1$, but $h(0) = \frac{1}{2}$. The function j is continuous at 0 since $\lim_{0} j = j(0) = 0$.

The function g is continuous at each number a in its domain by Definition 3.4. Although its graph consists of two separated pieces we choose nevertheless to call g a continuous function. The break in the graph is caused by a break in the domain which is not an interval.

DEFINITION 3.5 A function is *continuous* if it is continuous at each number in its domain.

Consider the functions k, l, m. See Figure 3.6.

$$\begin{aligned}
k\colon k(x) &= 1\ , && 0 \leqslant x < 1, \quad 2 < x \leqslant 3 \\
l\colon l(x) &= 1\ , && 0 \leqslant x < 1 \\
&= -1, && 1 < x \leqslant 2 \\
m\colon m(x) &= 1\ , && 0 \leqslant x \leqslant 1 \\
&= -1, && 1 < x \leqslant 2.
\end{aligned}$$

[9] The assertions $\lim_{a} f$ exists and $f(a)$ exists are implicit in this definition. If either f has no limit at a, or $a \notin \operatorname{dom} f$, or both, then f is not continuous at a.

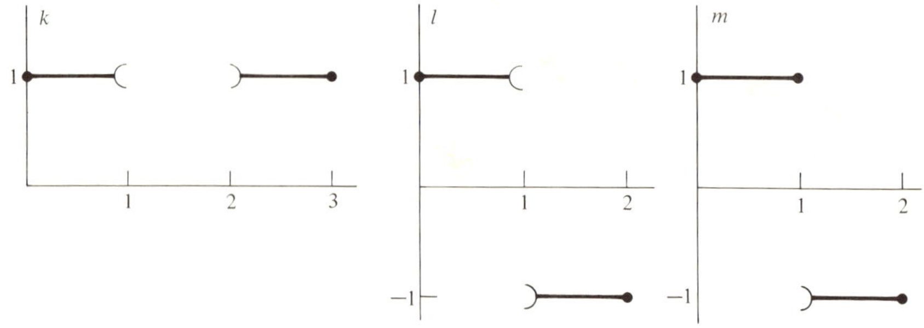

Figure 3.6

The functions k and l are continuous functions, but m is not a continuous function because at the number 1, in dom m, the limit $\lim_{1} m$ does not exist.

If the domain of a function contains an isolated point (see Section 3.1, page 72), the function is continuous at that point by Definition 3.4. Thus the function q,

$$\begin{aligned} q\colon q(x) &= 1 \quad , \quad x = \frac{1}{n}, \quad n = 1,2, \ldots, \\ &= \sqrt{x}, \quad 2n - 1 \leqslant x < 2n, \quad n = 1,2, \ldots, \end{aligned}$$

is continuous. See Figure 3.7. The literary connotation of the word continuous as "unbroken," is preserved whenever the domain of a function is an interval, for then the graph does not exhibit any gaps.

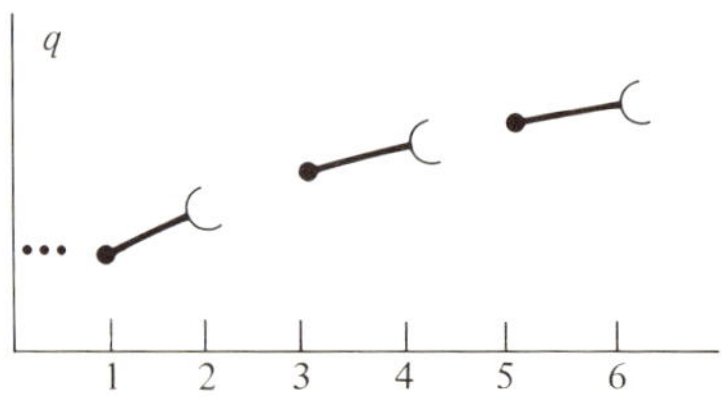

Figure 3.7

It is sufficient for most of the work in this book to assume that the domains of the functions we use are intervals, or at worst a union of intervals. For example, if we consider the reciprocal function r,

$$r\colon r(x) = \frac{1}{x}, \quad x \neq 0$$

at a point x_0 in its domain, there is an interval $|x - x_0| < \delta$, for $\delta > 0$, contained in dom r. Although r is a continuous function by Definition 3.5, whose graph has an infinite gap, yet on this δ-interval the values of r form a bounded set and the graph is unbroken. The function r_δ,

$$r_\delta\colon r_\delta(x) = r(x), \quad |x - x_0| < \delta$$

is called the *restriction* of r to the δ-interval.

Let us prove a few fundamental theorems which describe properties of continuous functions.

THEOREM 3.13 If the functions f and g are continuous, then so also are $f+g$, fg, and f/g, whenever these functions exist.[10]

Proof: If $a \in \text{dom}(f+g)$, then $a \in \text{dom } f$ and $a \in \text{dom } g$, and by hypothesis $\lim_a f = f(a)$, $\lim_a g = g(a)$. Therefore, by Theorem 3.7,

$$\lim_a (f+g) = \lim_a f + \lim_a g = f(a) + g(a) = (f+g)(a)$$

so that $f+g$ is continuous by Definition 3.5. Also by Theorem 3.8

$$\lim_a fg = (\lim_a f)(\lim_a g) = f(a)g(a) = (fg)(a)$$

and fg is continuous.

If $a \in \text{dom } f/g$, then $\lim_a g = g(a) \neq 0$ by Definition 3.2(b), and by Corollary to Theorem 3.9,

$$\lim_a \frac{f}{g} = \frac{\lim_a f}{\lim_a g} = \frac{f(a)}{g(a)} = \frac{f}{g}(a),$$

so that f/g is continuous. ■

It is clear now that since polynomial functions are continuous functions (see Section 3.4), rational functions are continuous also. In certain useful cases a number which is outside the domain of a rational function (hence at which it is not continuous) may still be an accumulation point of that domain, and as we observed in Section 3.4, the function may have a limit there which can be employed in place of the missing function value.[11]

The next theorem justifies the draftsman's practice of drawing graphs of continuous functions, whose domains are intervals, as unbroken curves. Read the theorem carefully, notice the hypothesis about the domain, and use the illustrative functions f, g, h, j, k, m, and q to help explain what it asserts.

THEOREM 3.14 Let f be a continuous function whose domain is an inter-

[10] This simply requires that $\text{dom } f \cap \text{dom } g \neq \varnothing$.

[11] You have noticed how sensitive these definitions and proofs are to the structure of the domain of a function. It is not surprising that great attention has been paid to point sets on a line. Such work is one aspect of point set topology.

val with numbers a and b in this domain. If $f(a) < f(b)$, and if K is a number between $f(a)$ and $f(b)$,

$$f(a) < K < f(b),$$

then there is a number c in the interval $[a,b]$ at which

$$f(c) = K.$$

Proof: Define the set M to be

$$M = \{x | f(x) < K, \quad x \in [a,b]\},$$

that is, the set of all numbers in $[a,b]$ at which f takes values less than K. See Figure 3.8. Since $f(a) < K$, $a \in M$, and M is not empty. Also, if $x \in M$, then $a \leqslant x \leqslant b$, and so the set M has the upper bound b. By the least upper bound axiom, the set M has a least upper bound, $c = \text{lub } M$. We shall show that $f(c) = K$.[12]

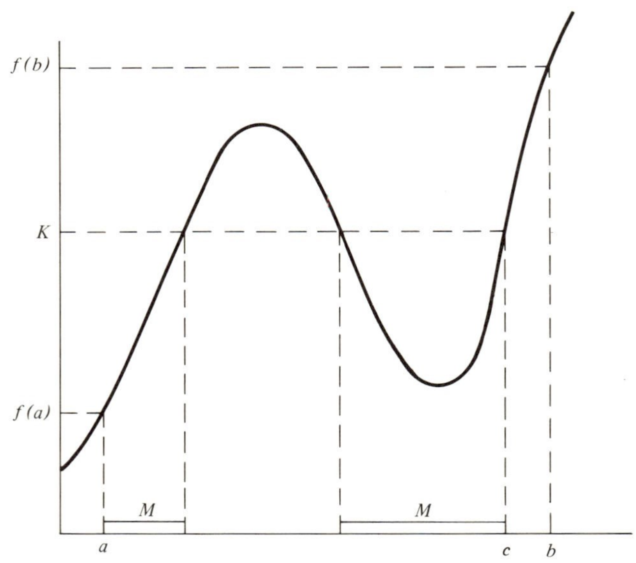

Figure 3.8

First, we locate c. Since $M \subset [a,b]$, there is an $m \in M$ such that

[12] Heretofore in this chapter we have typically assumed the existence of certain numbers we wanted to prove something about. Remember, we often have said "If $\lim_a f$ exists, then. . . ." In this case, though, we are up against proving the existence of c, and it takes the powerful least upper bound axiom to provide a candidate.

$a \leqslant m \leqslant c$; remember that $c = \text{ub } M$. See Figure 3.9. Also, $c \leqslant b$ by the definition of least upper bound. Thus, $a \leqslant c \leqslant b$, and $c \in \text{dom } f$.

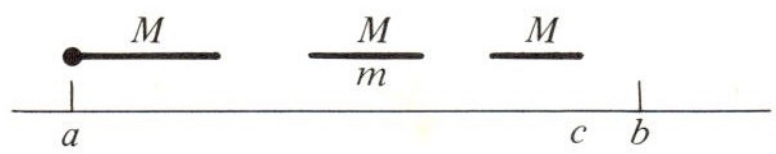

Figure 3.9

By the continuity of f, $\lim_c f = f(c)$; that is, for any $\epsilon > 0$, there is a $\delta > 0$ such that if $x \in [a,b]$ with $c - \delta < x < c + \delta$, then $f(x)$ is held in the interval

(A) $$f(c) - \epsilon < f(x) < f(c) + \epsilon.$$

Now $c - \delta$ is not an upper bound of M, because c is the least upper bound of M, so there is a number $x_0 \in M$ for which $f(x_0) < K$, with $c - \delta < x_0 \leqslant c$. See Figure 3.10. From (A), $f(c) - \epsilon < f(x_0) < K$, and this implies

(B) $$f(c) - \epsilon < K.$$

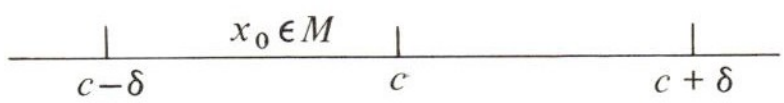

Figure 3.10

Possibly $c = b$. If so, $f(c) = f(b) > K$ by hypothesis. Hence, surely,

(C) $$K < f(c) + \epsilon.$$

Otherwise, $c < b$, and we take an x_0 in the interval $c < x_0 < c + \delta$, where $x_0 \notin M$, since $c = \text{ub } M$. Thus, $f(x_0) \not< K$, that is, $f(x_0) \geqslant K$. Again from (A), $K \leqslant f(x_0) < f(c) + \epsilon$, and (C) is true in all cases.

Combining (B) and (C) we get

(D) $$|f(c) - K| < \epsilon.$$

If $|f(c) - K| > 0$, we can take $\epsilon = \frac{1}{2}|f(c) - K|$, (remember ϵ is at our disposal), and (D) reads

$$|f(c) - K| < \tfrac{1}{2}|f(c) - K|,$$

a contradiction; therefore

$$|f(c) - K| = 0, \text{ that is, } f(c) = K. \blacksquare$$

3.5 PROBLEMS

1. Prove that if f and g are continuous functions, then so are both cf and $f - g$ (c is a real number).

2. Prove directly from Definitions 3.4 and 3.5 that each of the following functions is continuous.
   (a) $f(x) = x^2 + 2x - 1, \quad |x - 3| > 2$
   (b) $f(x) = x^2 + 2x - 1, \quad 0 < x \leqslant 1$
   (c) $f(x) = 1/x$
   (d) $f(x) = (x - 4)/(x + 2)$
   (e) $f(x) = |x + 3|, \quad x \neq -3$
3. Prove directly from Definitions 3.4 and 3.5 that none of the following functions is continuous.
   (a) $f(x) = x^2 + 2x - 1, \quad |x - 3| > 2, \quad x \neq -1$
   $f(-1) = 13$
   (b) $f(x) = 1/x, \quad x > 0$
   $= 4 \quad , \quad x = 0$
   (c) $f(x) = (x - 4)/(x + 2),$
   $f(-2) = 0$
   (d) $f(x) = |x + 3|, \quad x \neq -3$
   $f(-3) = -1$
4. Show that $f(x) = x^2 + 2x - 1$ is continuous at $x = 1$.
5. For each of the following functions, draw the graph and tell where the function does not have a limit and also where it is not continuous.
   (a) $f(x) = \dfrac{|x|}{2x}$
   (b) $f(x) = \begin{cases} |x|, & -1 \leqslant x \leqslant 0 \\ 2, & 0 < x \leqslant 1 \end{cases}$
   (c) $f(x) = \begin{cases} x, & -1 \leqslant x < 0 \\ 1, & x = 0 \\ 2, & 0 < x \leqslant 1 \end{cases}$
   (d) $f(x) = \begin{cases} x, & -1 \leqslant x < 0 \\ 1, & x = 0 \\ 2x, & 0 < x \leqslant 1 \end{cases}$
   (e) $f(x) = \begin{cases} -1, & -1 \leqslant x < 0 \\ x, & 0 \leqslant x < 1 \\ 2 - x, & 1 \leqslant x < 2 \\ 3, & x = 2 \\ x - 3, & 2 < x \leqslant 3 \end{cases}$
   (f) $f(x) = 1/x$
6. The domain of a function is to be the interval $[-2,+2]$. Specify such a function and draw its graph if
   (a) it has a limit at each point and is continuous at each point;
   (b) it has a limit at each point but is not continuous at $x = 0$ or at $x = 1$;
   (c) it has no limit when $x = 0$ and is not continuous when $x = 0,1,2$.

7. Let f be the function, dom $f = [0, 1]$, given by (see Problem 15, Section 2.6)
   (a) $f(0) = 1$
   (b) $f(x) = 0, \quad x$ irrational
   (c) $f(x) = 1/q, \quad x = p/q, \quad \text{g.c.d}(p,q) = 1$
   (d) $f(1) = 1$
   Prove that f is continuous at irrational numbers and is not continuous at rational numbers.
8. Use Problem 3(a) to show that continuity is not a necessary condition that the number c of Theorem 3.14 exist for some interval $[a,b]$. Take $a = -7$, $b = 6$ so that $f(a) = 34 < 47 = f(b)$. Find c for $K = 40$.
9. For each of the continuous functions in Problem 2 choose a and b in some interval contained in dom f, and a number K between $f(a)$ and $f(b)$. Find a corresponding number c.
10. We have seen that the function f, where

$$f(x) = x \sin \frac{1}{x}, \quad x \neq 0$$
$$f(0) = 0$$

is a continuous function. Let $a = (1/5\pi)$, $b = (2/\pi)$. Then

$$f(a) = 0 < \frac{2}{\pi} = f(b).$$

Find all values of $c \in (a,b)$ at which $f(c) = 2/9\pi$. Show that for any positive integer n there is an interval $[a,b] \subset$ dom f, where $f(a) < f(b)$, such that for some K between $f(a)$ and $f(b)$, $f(c) = K$ for n distinct values of c in (a,b).
11. State and prove a theorem like Theorem 3.14 with the inequalities reversed.
12. (a) Prove that if P is a polynomial function and either $P(a) < 0 < P(b)$ or $P(b) < 0 < P(a)$ for $a < b$, then the equation $P(x) = 0$ has a root in (a,b).
    (b) Give an example of a polynomial function such that $P(x) \geqslant 0$ for all x, and $P(x) = 0$ has a root. (An application of (a) will not necessarily catch all roots.)
13. Show that the equation

$$4x^3 - 9x^2 - 6x + 2 = 0$$

has a root in each of the intervals $(-1,0)$, $(0,1)$, and $(2,3)$.
14. Let f be given by

$$f(x) = \frac{1}{x}, \quad x \neq 0$$
$$f(0) = 1.$$

Notice that $f(-1) = -1 < 1 = f(1)$, and yet $f(x) = 0$ has no root. Is this a counterexample to Theorem 3.14?

3.6 CONTINUITY OF THE COMPOSITION

We have seen that the sum, product, and quotient of continuous functions are each continuous. Turning to compositions, consider the example,

$$\begin{aligned} f&: f(y) = \sqrt{y} \quad, && y \geqslant 0 \\ g&: g(x) = x^2 - 1, && x \text{ real}, \end{aligned}$$

where each of f and g is a continuous function. The composition $f \circ g$ is

$$f \circ g\colon (f \circ g)(x) = \sqrt{x^2 - 1}, \quad x \leqslant -1, \quad x \geqslant 1$$

whose domain is not an interval, and our next theorem will demonstrate that it is continuous.

Let f and g be the functions.

$$\begin{aligned} f&: f(y) = \sqrt{y} \quad, && y \geqslant 0 \\ g&: g(x) = x^2(x - 1), && x \text{ real}. \end{aligned}$$

Notice that $g(x)$ is positive for $x > 1$, is zero at $x = 0,1$, and is negative elsewhere. So the composition

$$(f \circ g)(x) = |x| \sqrt{x - 1}, \quad x = 0, \quad x \geqslant 1$$

has the isolated point 0 in its domain. As we observed in Section 3.5, the composition $f \circ g$ is continuous at $x = 0$.

THEOREM 3.15 Let f and g be continuous functions with the composition $f \circ g$. Then $f \circ g$ is continuous.

Proof: Choose any number $a \in \operatorname{dom} f \circ g$; then $g(a) \in \operatorname{dom} f$. Since f is continuous, $\lim_{g(a)} f = f(g(a)) = (f \circ g)(a)$, so by Definition 3.2, for any $\epsilon > 0$ there is a $\delta_1 > 0$ such that for $y \in \operatorname{dom} f$ and $|y - g(a)| < \delta_1$ it is true that

$$|f(y) - f(g(a))| = |f(y) - (f \circ g)(a)| < \epsilon.$$

With this number δ_1, and the continuity of g, $\lim_{a} g = g(a)$, we know there is a $\delta > 0$ so that for $x \in \operatorname{dom} g$ and $|x - a| < \delta$, it is true that $|g(x) - g(a)| < \delta_1$.

Put this together as follows. Choose $\epsilon > 0$ and take the number δ

defined above. Then if $x \in \operatorname{dom} f \circ g$ with $|x - a| < \delta$, it is true that $g(x) \in \operatorname{dom} f$ with $|g(x) - g(a)| < \delta_1$; so therefore

$$|(f \circ g)(x) - (f \circ g)(a)| = |f(g(x)) - f(g(a))| < \epsilon. \blacksquare$$

3.6 PROBLEMS

1. In Problem 8, page 61, for each of the parts in which f and g are continuous functions, verify that $f \circ g$ is continuous.
2. Let f and g be given by

$$\begin{aligned} f(x) &= x \sin \frac{1}{x}, && x \neq 0 \\ f(0) &= 0 &&, \\ g(x) &= \frac{1}{x} &&, \quad x > 0 \end{aligned}$$

   so that both are continuous functions. Examine the continuity of $f \circ g$ and $g \circ f$.
3. Let f and g be given by

$$\begin{aligned} f(x) &= \sqrt{x} &&, \quad x \geqq 0 \\ g(x) &= 2x^3 - x^2 - x^4, && x \geqq 0. \end{aligned}$$

   Examine the continuity of $f \circ g$.

An important application of our study of the composition of functions can be made to the relation between a function and its inverse function. It is true for any function

$$f\colon x \longrightarrow f(x), \quad x \in \operatorname{dom} f$$

that $x_1 = x_2$, in dom f, implies $f(x_1) = f(x_2)$ in range f. If it should happen that $f(x_1) = f(x_2)$ in range f implies $x_1 = x_2$ in dom f, then each number in range f has a unique pre-image[13] in dom f, and it is evident that the relation f^{-1},

$$\text{(A)} \qquad f^{-1}\colon f(x) \longrightarrow x = f^{-1}(f(x)), \quad f(x) \in \operatorname{range} f$$

is a function whose domain is range f.

DEFINITION 3.6 If the function f has the property $f(x_1) = f(x_2)$ implies $x_1 = x_2$, then the function f^{-1} defined by (A) is the *inverse* of f.

[13] Just as $f(x)$ is called the *image* of x in f, so is x called the *pre-image* of $f(x)$ in f.

THEOREM 3.16 If a function f has an inverse, f^{-1}, then[14]

$$f^{-1} \circ f = I_d \qquad \text{and} \qquad f \circ f^{-1} = I_r,$$

where I_d is the identity function with dom $I_d =$ dom f, and I_r is the identity function with dom $I_r =$ range f.

Proof: If $x \in$ dom f, then $f(x) \in$ range f, and from (A) $(f^{-1} \circ f)(x) = f^{-1}(f(x)) = x$, $x \in$ dom f, and therefore $f^{-1} \circ f = I_d$.

Suppose that $y \in$ range f; then $y = f(x)$ for some single $x \in$ dom f and by (A),

$$\begin{aligned}(f \circ f^{-1})(y) &= f(f^{-1}(y)) = f(f^{-1}(f(x))) \\ &= f(x) = y,\end{aligned}$$

that is, $f \circ f^{-1} = I_r$. ■

We note in passing that $(f^{-1})^{-1} = f$, which you can verify easily.

A sufficient condition for a function f to possess an inverse is given following the definition.

DEFINITION 3.7 A function is *monotone increasing* if for each x_1 and x_2 in dom f, with $x_1 < x_2$, it is true that $f(x_1) \leqslant f(x_2)$. A function f is *strictly monotone increasing* if under these conditions $f(x_1) < f(x_2)$. The terms *monotone decreasing* and *strictly monotone decreasing* are defined similarly.

THEOREM 3.17 If a function f is strictly monotone increasing (decreasing), then f has an inverse, f^{-1}, which is also strictly monotone increasing (decreasing).

Proof: We prove the increasing case. Consider the rule as in (A):

$$f^{-1}: f(x) \longrightarrow x = f^{-1}(f(x)), \quad f(x) \in \text{range } f.$$

Take dom $f^{-1} =$ range f. We now verify that the rule given assigns a unique image to each number $f(x) \in$ dom f^{-1}. Suppose that $f(x_1) = f(x_2)$. If $x_1 < x_2$ or $x_2 < x_1$, then by the monotonicity of f either $f(x_1) < f(x_2)$ or $f(x_2) < f(x_1)$—both false. Therefore, $x_1 = x_2$ and

$$f^{-1}(f(x_1)) = x_1 = x_2 = f^{-1}(f(x_2)),$$

and f^{-1} is a function.

If $f(x_1) < f(x_2)$, then $x_2 \nleqslant x_1$ by the monotonicity of f; therefore, $x_1 < x_2$, and

[14] Any identity function has the rule $I(x) = x$.

$$f^{-1}(f(x_1)) = x_1 < x_2 = f^{-1}(f(x_2)). \blacksquare$$

Lemma to Theorem 3.18: If a continuous function f, whose domain is an interval, has an inverse, then f is strictly monotone, either increasing or decreasing.

Proof: Clearly, if $x_1 \neq x_2$, then $f(x_1) \neq f(x_2)$. So if we show monotonicity, it will be of the strict kind. If f is not decreasing, then either it is not monotone or it is increasing. If f is not monotone, then there are numbers $x_1 < x_2 < x_3$ so that either

$$f(x_1) < f(x_2) > f(x_3)$$

as in Figure 3.11 or

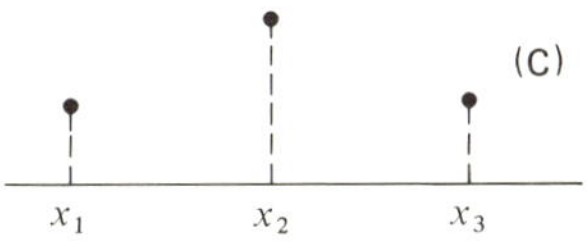

Figure 3.11

$$f(x_1) > f(x_2) < f(x_3)$$

as in Figure 3.12.

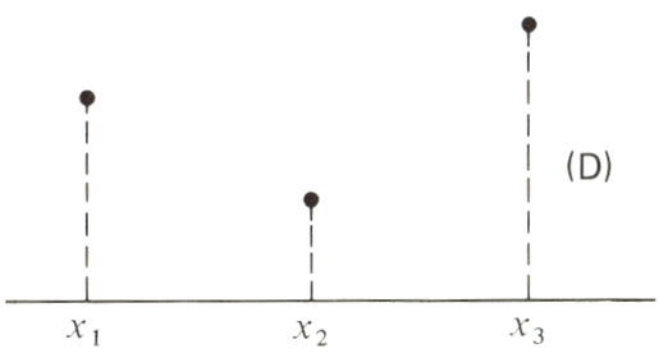

Figure 3.12

Suppose that (C) were the case. Choose K so that $f(x_1) < K < f(x_2)$, $f(x_2) > K > f(x_3)$. Then[15] by Theorem 3.14 there would be numbers c_1 and c_2 so that

$$f(c_1) = K, \quad x_1 < c_1 < x_2$$

and

$$f(c_2) = K, \quad x_2 < c_2 < x_3.$$

That is,

[15] Recall that Theorem 3.14 rests on the least upper bound axiom. Without that axiom a discussion of a continuous function is very awkward.

$$f(c_1) = f(c_2), \quad c_1 \neq c_2,$$

contradicting the existence of f^{-1}. Similarly, case (D) cannot occur; therefore, f is strictly monotone increasing. See Figure 3.13.

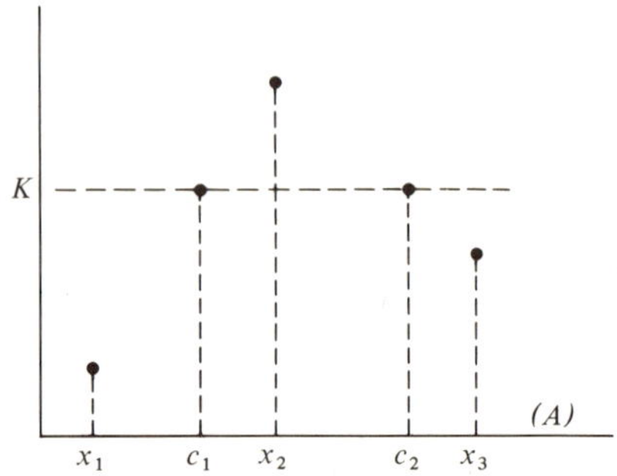

Figure 3.13

By the same kind of argument, if f is not increasing, it is strictly monotone decreasing. ■

THEOREM 3.18 Let a continuous function f, whose domain is an interval, have the inverse f^{-1}. Then f^{-1} is continuous also.

Proof: Let $y_0 \in \operatorname{dom} f^{-1}$ so that $f^{-1}(y_0)$ exists in dom f and by the lemma we may suppose that f is increasing. Choose $\epsilon > 0$. Then there is an x_1, and/or an x_2 in dom f satisfying $f^{-1}(y_0) - \epsilon < x_1 < f^{-1}(y_0) < x_2 < f^{-1}(y_0) + \epsilon$. Take $\delta > 0$ so that $\delta = \min[(f(x_2) - y_0), (y_0 - f(x_1))]$. Then if $y \in \operatorname{dom} f^{-1}$ with $|y - y_0| < \delta$, it follows that (see Figure 3.14)

$$f(x_1) \leq y_0 - \delta < y < y_0 + \delta \leq f(x_2).$$

Since f is monotone (we are doing the increasing case), f^{-1} is monotone increasing by Theorem 3.17, and

$$f^{-1}(y_0) - \epsilon < x_1 < f^{-1}(y) < x_2 < f^{-1}(y_0) + \epsilon,$$

which, in the symbolism of the Definition 3.2 of a limit, is

$$|f^{-1}(y) - f^{-1}(y_0)| < \epsilon.$$

The argument for the case f decreasing is symmetric to the one just given.[16] ■

[16] We repeat: Observe how critical it is to have complete information about the domains of functions.

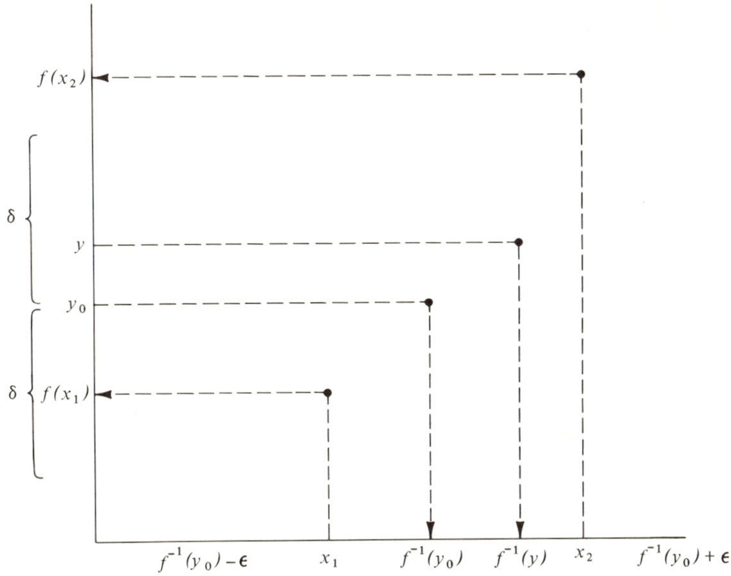

Figure 3.14

3.6 PROBLEMS (*continued*)

4. Examine each of the following functions for possession of an inverse, based on Definition 3.6.
   (a) $f(x) = x^2$,
   (b) $f(x) = x^2, \quad 0 \leqslant x < 10$
   (c) $f(x) = x^2, \quad -1 < x < 10$
   (d) $f(x) = x^2, \quad -1 \leqslant x \leqslant 0, \quad x > 1$
5. Which of the functions in Problem 4 are monotone? continuous?
6. Define I_d and I_r for each of those functions in Problem 4 which has an inverse. Verify in each case that $(f^{-1})^{-1} = f$.
7. Show by an example that monotonicity is not a necessary condition for the existence of an inverse.
8. Write a proof of Theorem 3.17 for the decreasing case.
9. Show by a counterexample that the Lemma to Theorem 3.18 would be false if the hypothesis of the continuity of f were removed. Complete the proof of the lemma, as stated.
10. Show by an example that there exists a function with an inverse which is not continuous.
11. Complete the proof of Theorem 3.18 by writing out the argument for the decreasing case.
12. (a) Explain why the function sine has no inverse.

(b) Define a function $\sin_I$ by

$$\sin_I x = \sin x, \quad I \text{ is an interval and } x \in I.$$

Which of the following functions has an inverse?

(i) $\sin_I$, $I = \{x | 0 \leqslant x < \pi\}$
(ii) $\sin_I$, $I = \{x | -\pi \leqslant x \leqslant -\pi/2\}$
(iii) $\sin_I$, $I = \{x | 5\pi/2 \leqslant x \leqslant 7\pi/2\}$
(iv) $\sin_I$, $I = \{x | -\pi/2 \leqslant x \leqslant \pi/2\}$

In the literature the notation "arcsin" is used for $\sin_I^{-1}$ in case (iv).

Chapter 4
THE DERIVATIVE FUNCTION

4.1 BASIC DEFINITIONS

Our discussion of continuity was generated by a desire to formulate, in the class of functions, a property which would correspond to "unbrokenness" in the class of geometric graphs of these functions. We discovered that at worst the graph of a continuous function consists of separated points and separated pieces of unbroken curves. This geometric motivation produced a sequence of interesting theorems about functions, so we are encouraged to try the geometric route again.

In Section 2.6 we looked at a function whose graph, though unbroken, contained a corner. The slope function appeared to be sensitive to the existence of such a corner, and it is this observation that we proceed to develop systematically now. The geometric idea involved here is "smoothness," that is, the absence of abrupt change of direction, as at a corner.

The essential tool used in this case is the slope function described by the next definition.

DEFINITION 4.1 Let f be a function, and $x \in \operatorname{dom} f$. The function

$$_f\sigma_x\colon\ _f\sigma_x(h) = \frac{1}{h}\,[f(x+h) - f(x)], \quad \{h \mid h \neq 0, \quad x+h \in \operatorname{dom} f\}$$

is the *slope function* for f at x.

As we observed before in Sample 4, Section 2.6, a value of the slope function measures the slope of a secant line joining the point $P(x,f(x))$ on the graph of f to a nearby point $Q(x+h,f(x+h))$ on that graph. See Figure 4.1. At each number $x \in \operatorname{dom} f$ there is associated a particular slope function $_f\sigma_x$, which is always discontinuous at 0 (since 0 is not in its domain) but which may have a limit at 0.

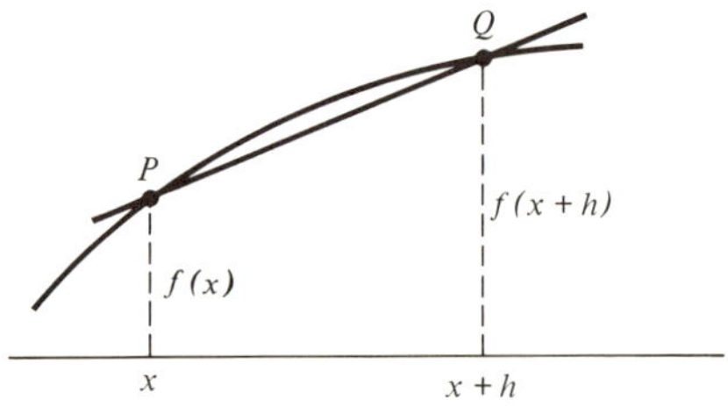

Figure 4.1

DEFINITION 4.2 For a function f and a number $x \in \operatorname{dom} f$, if $\lim\limits_{0}\ _f\sigma_x$ exists, it is called the *slope of the graph* of f at x.

In case x is an isolated point in dom f, this limit will not exist. Therefore, the graph of a function f can have a slope only at a number which is in dom f and not isolated from dom f.

With this background we translate geometrical smoothness of a graph into *differentiability* of the corresponding function by two more definitions.

DEFINITION 4.3 Let f be a function whose domain is an interval. If $\lim\limits_{0}\ _f\sigma_x$ exists, then f is *differentiable at* x. If $\lim\limits_{0}\ _f\sigma_x$ exists for each $x \in \operatorname{dom} f$, then f is a *differentiable* function.

By Definition 4.3, if a function f is differentiable, then to each $x \in \operatorname{dom} f$ (an interval) there corresponds a unique limit of the slope function. This justifies the last definition on this list.

DEFINITION 4.4 If f is differentiable, the function[1]

$$Df\colon Df(x) = \lim_{0} {}_f\sigma_x, \quad x \in \operatorname{dom} f$$

is called the *derivative of f*.

Let us examine some examples which make use of these definitions.

EXAMPLE 1

$$\begin{aligned} f(x) &= 4x - x^2, \quad 0 \leqslant x \leqslant 1 \\ {}_f\sigma_x(h) &= \frac{1}{h}\left[(4(x+h) - (x+h)^2) - (4x - x^2)\right] \\ &= 4 - 2x - h, \quad x \in \operatorname{dom} f, \quad h \neq 0 \\ Df(x) &= \lim_{0} {}_f\sigma_x = 4 - 2x, \quad x \in \operatorname{dom} f \end{aligned}$$

In Figure 4.2 the slope of the graph of f is 4 at $x = 0$, is 3 at $x = \frac{1}{2}$, and is 2 at $x = 1$. The function f is differentiable at each number x in its domain; hence, f is differentiable and its derivative is $Df(x) = 4 - 2x$, $0 \leqslant x \leqslant 1$. You should sketch a graph of f, several graphs of ${}_f\sigma_x$ (for different values of x), and one of Df. Notice that each ${}_f\sigma_x$ is discontinuous at zero.

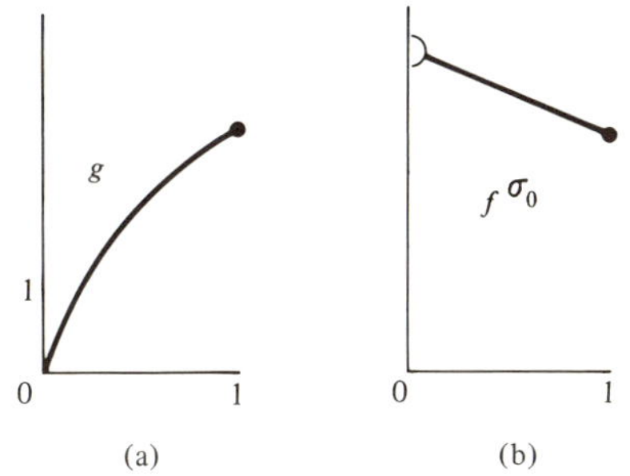

Figure 4.2

EXAMPLE 2

$$\begin{aligned} g(x) &= 4x - x^2, \quad 0 \leqslant x \leqslant 1 \\ &= 6 - 3x, \quad 1 < x \leqslant 2 \end{aligned}$$

Notice that g is a continuous function. (See Figure 4.3.) For $-x \leqslant h \leqslant 2 - x$,

$$\begin{aligned} {}_g\sigma_x(h) &= 4 - 2x - h, \quad 0 \leqslant x < 1, \quad h \neq 0 \\ {}_g\sigma_1(h) &= 2 - h, \quad h < 0 \\ &= -3, \quad h > 0 \end{aligned}$$

[1] The symbol $Df(x)$ is an abbreviation for the symbol $(Df)(x)$.

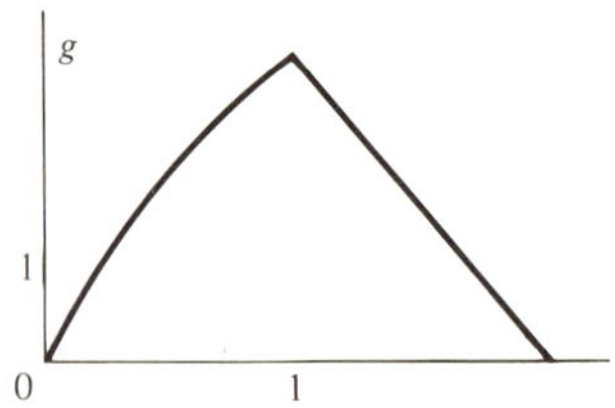

Figure 4.3

$$_g\sigma_x(h) = -3 \qquad , \quad 1 < x \leq 2, \quad h \neq 0.$$

(See Figure 4.4.)

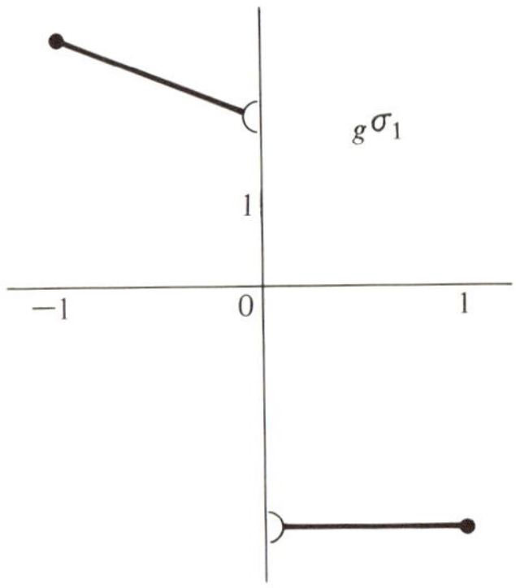

Figure 4.4

In this case

$$\begin{aligned} &\lim_0 {}_g\sigma_x = 4 - 2x, \quad 0 \leq x < 1 \\ &\lim_0 {}_g\sigma_1 \quad \text{does not exist} \\ &\lim_0 {}_g\sigma_x = -3 \quad , \quad 1 < x \leq 2. \end{aligned}$$

Therefore, g is not a differentiable function, since the slope function at $x = 1$ has no limit at 0. A graph of g shows a corner at $x = 1$.

EXAMPLE 3 $\quad j(x) = 4x - x^2, \quad 0 \leq x < 1$
$\qquad\qquad = 6 - 3x \;, \quad 1 < x \leq 2$

We have deleted the number $x = 1$ from dom g to get j, and the computations are already done in Example 2. Thus j is a differentiable function, and its derivative Dj is

$$\begin{aligned} Dj(x) &= 4 - 2x, \quad 0 \leq x < 1 \\ &= -3 \quad , \quad 1 < x \leq 2. \end{aligned}$$

EXAMPLE 4 (See Section 3.2.) Let a function f be given by

$$f\colon f(t) = 16t^2, \quad 0 \leqslant t \leqslant 4.$$

Previously we computed

$$_f\sigma_t(h) = 32t + 16h, \quad 0 \leqslant t \leqslant 4, \quad h \neq 0$$

so that f is a differentiable function, and has the derivative

$$Df\colon Df(t) = 32t, \quad 0 \leqslant t \leqslant 4.$$

In the mechanical context f is a displacement function, and Df is the corresponding velocity function. It is said that the velocity is proportional to the time elapsed from rest. It is easy to continue one more step and compute

$$\begin{aligned} {}_{Df}\sigma_t(h) &= \frac{1}{h}\,[32(t+h) - 32t] \\ &= 32, \quad 0 \leqslant t \leqslant 4, \quad h \neq 0, \end{aligned}$$

so that Df is itself a differentiable function with the derivative[2]

$$D^2f\colon D^2f(t) - 32, \quad 0 \leqslant t \leqslant 4.$$

This, D^2f, is called the *second derivative* of f, and measures the acceleration of the particle which, under the displacement hypothesis, is constant.[3]

4.1 PROBLEMS

For each of the functions in Problems 1 and 2 find those numbers at which the function is differentiable; that is, at which $\lim\limits_{0} {}_f\sigma_x$ exists. Classify each function as not differentiable or differentiable.

1. (a) $f(x) = 3x^2$
   (b) $f(x) = 2x - 3$
   (c) $f(x) = x^3$
   (d) $f(x) = 1 - x^2$
   (e) $f(x) = 2x - x^2$
   (f) $f(x) = 1/x, \quad x \neq 0$
   (g) $f(x) = x/(1 + x), \quad x \neq -1$
   (h) $f(x) = \sqrt{x}, \quad x \geqslant 0$
2. (a) $f(x) = \sqrt{1 - x^2}, \quad |x| \leqslant 1$

[2] D^2f is an abbreviation for $D(Df)$.

[3] It is more usual, in physics, to take the constant acceleration function as an hypothesis, and to deduce the quadratic displacement function as a consequence. You can undertake this later by using Theorem 5.2 and its corollary.

(b) $f(x) = |x|$
(c) $f(x) = |x - 2|$
(d) $f(x) = x|x|$
(e) $f(x) = |1 - x^2|$
(f) $f(x) = |x^3|$

3. For each function f in Problems 1 and 2 which is differentiable, write a formula for the derivative Df.
4. One of the functions f, g, h given below is differentiable. Write a formula for its derivative.

$$\begin{aligned} f(x) &= \sin \pi/x \quad, & x \neq 0 \\ &= 0 \quad, & x = 0 \\ g(x) &= x \sin \pi/x \,, & x \neq 0 \\ &= 0 \quad, & x = 0 \\ h(x) &= x^2 \sin \pi/x, & x \neq 0 \\ &= 0 \quad, & x = 0 \end{aligned}$$

5. In each case below a rule is given. Adjoin to this rule a domain so that the resulting function f has a differentiable derivative. Write a formula for D^2f.
(a) $f(x) = 3(x/2)$
(b) $f(x) = x^2$
(c) $f(x) = x^5$
(d) $f(x) = 1/x$
(e) $f(x) = \sqrt{x}$
(f) $f(x) = \sqrt{1 - x}$
6. Find D^3f for each of the following.
(a) $f(x) = 4 - 7x$
(b) $f(x) = x^2 + 2x$
(c) $f(x) = x^3$
(d) $f(x) = 2x - 2$
(e) $f(x) = x^3 - 3x^2 - 9x + 9$

4.2 THEOREMS ABOUT THE DERIVATIVE FUNCTION

The examples of the last section were devoted to the computation of derivatives of certain special functions. It was necessary to examine the slope function in each case since all we had to work with were the basic definitions in which the slope function was involved. In the study of continuity in the last chapter some general theorems were proved which released us from direct dependence on the definitions and ultimately made the work easier. So, also, is it with differentiability; some theorems will be helpful.

THEOREM 4.1 If f is a constant function whose domain is an interval, then f is differentiable and Df is a 0 constant function.[4]

Proof: Let f be the function

$$f\colon f(x) = c, \quad a \leqslant x \leqslant b, \quad c \text{ is a real number.}$$

Then

$$\begin{aligned} {}_f\sigma_x(h) &= \frac{1}{h}\,[f(x+h) - f(x)] \\ &= \frac{1}{h}\,[c - c] \\ &= 0, \quad a \leqslant x \leqslant b, \quad h \neq 0, \end{aligned}$$

and therefore

$$Df(x) = \lim_{0}\, {}_f\sigma_x = 0, \quad a \leqslant x \leqslant b. \blacksquare$$

THEOREM 4.2 If f is an identity function whose domain is an interval, then f is differentiable and Df is the constant function whose range contains the single number 1.

Proof: Let f be the function

$$f\colon f(x) = x, \quad a \leqslant x \leqslant b.$$

Then

$$\begin{aligned} {}_f\sigma_x(h) &= \frac{1}{h}\,[f(x+h) - f(x)] \\ &= \frac{1}{h}\,[x + h - x] \\ &= 1, \quad a \leqslant x \leqslant b, \quad h \neq 0, \end{aligned}$$

and therefore

$$Df(x) = \lim_{0}\, {}_f\sigma_x = 1, \quad a \leqslant x \leqslant b. \blacksquare$$

THEOREM 4.3 If f and g are differentiable functions, then $f + g$ is differentiable,[5] and $D(f + g) = Df + Dg$.

Proof: Since the domains of f and g are intervals (see Definition 4.3), the domain of $f + g$ is an interval, say $[a,b]$. Then

[4] See page 140 for a restricted converse, Theorem 5.2.

[5] We will assume that dom f and dom g overlap in at least two points.

$$\begin{aligned} {}_{f+g}\sigma_x(h) &= \frac{1}{h}\,[(f+g)(x+h) - (f+g)(x)] \\ &= \frac{1}{h}\,[(f(x+h) - f(x)) + (g(x+h) - g(x))] \\ &= {}_f\sigma_x(h) + {}_g\sigma_x(h), \quad a \leqslant x \leqslant b, \quad h \neq 0, \end{aligned}$$

and therefore

$$\begin{aligned} D(f+g)(x) &= \lim_0 {}_{f+g}\sigma_x \\ &= \lim_0 {}_f\sigma_x + \lim_0 {}_g\sigma_x \\ &= Df(x) + Dg(x), \quad a \leqslant x \leqslant b. \ \blacksquare \end{aligned}$$

THEOREM 4.4 (See the definition of cf, page 55.) If f is a differentiable function and c is a number, then cf is differentiable and $D(cf) = cDf$.

Proof: Since $\mathrm{dom}(cf) = \mathrm{dom}\, f$, an interval, and

$$\begin{aligned} {}_{cf}\sigma_x(h) &= \frac{1}{h}\,[(cf)(x+h) - (cf)(x)] \\ &= \frac{c}{h}\,[f(x+h) - f(x)] \\ &= c_f\sigma_x(h), \quad x \in \mathrm{dom}\, f, \quad h \neq 0, \end{aligned}$$

then

$$\begin{aligned} D(cf)(x) &= \lim_0 {}_{cf}\sigma_x \\ &= c \lim_0 {}_f\sigma_x \\ &= cDf(x). \ \blacksquare \end{aligned}$$

Corollary to Theorem 4.4: If f and g are differentiable functions, then $(f - g)$ is differentiable, and $D(f - g) = Df - Dg$.

Proof: By Theorem 4.4, $D(-g) = D((-1)g) = (-1)Dg = -Dg$. Therefore, Theorem 4.3 gives

$$D(f-g) = D(f + (-g)) = Df + D(-g) = Df - Dg. \ \blacksquare$$

We state and prove a lemma now, because it is useful in the proof of the next theorem. However, this lemma holds interest in its own right since it establishes that the class of differentiable functions is a subclass of the class of continuous functions.

Lemma to Theorem 4.5 If a function f is differentiable, then it is continuous.

Proof: By hypothesis, $\mathrm{dom}\, f$ is an interval, and $\lim_0 {}_f\sigma_x = Df(x)$ for $x \in \mathrm{dom}\, f$. For any $\epsilon > 0$, there is a $\delta_1 > 0$ so that if $0 < |h| < \delta_1$, then

$$\left|\frac{1}{|h|}|f(x+h)-f(x)|-|Df(x)|\right| \leq |{}_f\sigma_x(h)-Df(x)| < \epsilon,$$

the first inequality being an application of the triangle inequality. It follows that

$$\text{(A)} \qquad |f(x+h)-f(x)| < |h|(|Df(x)|+\epsilon).$$

Define the number $\delta = \min\left(\delta_1, \frac{\epsilon}{|Df(x)|+\epsilon}\right)$, then if $0 < |h| < \delta$, the inequality (A) will still be true, and

$$|f(x+h)-f(x)| < \delta(|Df(x)|+\epsilon) \leq \epsilon.$$

Thus, $\lim_{h\to 0} f(x+h) = f(x)$, for $x \in \operatorname{dom} f$, and f is a continuous function.[6] ■

THEOREM 4.5 If f and g are differentiable functions, then fg is differentiable, and $D(fg) = fDg + gDf$.

Proof: With the assumption of the footnote to Theorem 4.3, we see that dom fg is an interval. Furthermore,

$$\begin{aligned}
{}_{fg}\sigma_x(h) &= \frac{1}{h}[(fg)(x+h)-(fg)(x)] \\
&= \frac{1}{h}[f(x+h)g(x+h)-f(x)g(x)] \\
&= \frac{1}{h}[f(x+h)[g(x+h)-g(x)]+[f(x+h)-f(x)]g(x)] \\
&= f(x+h)\,{}_g\sigma_x(h) + g(x)\,{}_f\sigma_x(h), \quad x \in \operatorname{dom} fg.
\end{aligned}$$

By the lemma, f is continuous; that is, $\lim_{h\to 0} f(x+h) = f(x)$, and therefore

$$\lim_0 {}_{fg}\sigma_x = f(x)\lim_0 {}_g\sigma_x + g(x)\lim_0 {}_f\sigma_x, \quad x \in \operatorname{dom} fg,$$

that is,

$$D(fg) = fDg + gDf. \;■$$

We can easily arrive at a conclusion about the differentiability of the quotient of functions and of the inverse of a function when we know the next theorem about the differentiability of the composition of differentiable functions.

[6] The fact that f is continuous at x was previously defined by $\lim_x f = f(x)$. Show, as an exercise, that this is, in fact, what we have proved.

Lemma to Theorem 4.6:[7] Let the functions f and g have the composition $f \circ g$. If $\lim_a g = G$, $G \notin \text{dom } f$ and $\lim_G f = F$, then $\lim_a f \circ g = F$.

Proof: By hypothesis, for any $\epsilon > 0$ there is a $\delta_1 > 0$ so that for each $y \in \text{dom } f$ satisfying $0 < |y - G| < \delta_1$, the inequality $|f(y) - F| < \epsilon$ holds. Also, whatever δ_1 is, there is a $\delta > 0$ so that for each $x \in \text{dom } g$ satisfying $0 < |x - a| < \delta$, the inequality $|g(x) - G| < \delta_1$ is true. Therefore, for any $\epsilon > 0$, take the number δ given just above, and for each $x \in \text{dom } f \circ g$ (notice that $g(x) \neq G$; otherwise $G \in \text{dom } f$) and in $0 < |x - a| < \delta$, $g(x)$ will be in the interval $|g(x) - G| < \delta_1$, and $g(x) \neq G$. Therefore, call $g(x) = y$ and apply the first sentence of this proof; that is, $|f \circ g(x) - F| = |f(g(x)) - F| < \epsilon$ and so $\lim_a f \circ g = F$. ■

THEOREM 4.6 (chain rule) Let f and g be differentiable functions with the composition $f \circ g$ whose domain is an interval. Then $f \circ g$ is differentiable, and

$$D(f \circ g) = ((Df) \circ g)\, Dg.$$

Proof: The slope function for the composition is

$$_{f\circ g}\sigma_x(h) = \frac{1}{h}\,[f(g(x+h)) - f(g(x))], \quad h \neq 0,\ x + h \in \text{dom } f \circ g.$$

We show the slope functions for f and g as follows: Define a function k by[8]

$$k(h) = g(x+h) - g(x), \quad x + h \in \text{dom } g.$$

Then

$$_{f\circ g}\sigma_x(h) = \frac{1}{h}\,[f(g(x) + k(h)) - f(g(x))], \quad h \neq 0,$$

If there is a deleted interval I at 0 on which $k(h) \neq 0$, then

$$\text{(B)} \qquad \begin{aligned} _{f\circ g}\sigma_x(h) &= \frac{1}{k(h)}\,[f(g(x) + k(h)) - f(g(x))]\,\frac{k(h)}{h} \\ &= ({}_f\sigma_{g(x)} \circ k)\ {}_g\sigma_x(h), \quad h \neq 0, \quad h \in I. \end{aligned}$$

By the lemma, $\lim_0 {}_f\sigma_{g(x)} \circ k = Df(g(x))$, and therefore

[7] Compare this lemma with Theorem 3.15 (page 92). If f and g are both continuous functions, then $G = \lim_a g = g(a)$, so $G \in \text{dom } f$ and $\lim_a f \circ g = f(g(a)) = f(G) = \lim_G f = F$, by Theorem 3.15. However, for Theorem 4.6 we will need this result in case $G = 0$ and f is a slope function which is not continuous at 0 (see line (B) in the proof of Theorem 4.6). Therefore, Theorem 3.15 is not strong enough for our purpose here.

[8] To each number x there is associated such a function $_xk$. We omit the subscript for simplicity.

$$D(f \circ g)(x) = Df(g(x)) \lim_{0} {}_g\sigma_x$$
$$= ((Df) \circ g)(x)\, Dg(x), \quad x \in \operatorname{dom} f \circ g$$

as desired.

If there is no such interval I on which $g(x + h) \neq g(x)$, we are in trouble with the denominator $k(h)$ at line (B). In order to get rid of this trouble, consider the function[9] ψ,

$$\begin{aligned} \psi(h) &= ({}_f\sigma_{g(x)} \circ k)(h) - Df(g(x)), \quad & k(h) \neq 0 \\ &= 0 \quad &, \quad k(h) = 0. \end{aligned}$$

Then the slope function for the composition is

$$_{f\circ g}\sigma_x(h) = [Df(g(x)) + \psi(h)]\ {}_g\sigma_x(h), \quad h \neq 0.$$

Notice that this reduces to $0 = 0$ in case $k(h) = 0$. Again by the lemma, $\lim_{0} \psi = 0$, so the slope function for the composition has a limit at 0. ■

4.2 PROBLEMS

The problems in this list anticipate some of the discussion in Section 4.3. This should make your reading easier. In each problem make a note of the theorems you use.

1. Find Df if
   (a) $f(x) = x^3 + x^2 + x + 1$
   (b) $f(x) = 3x^2 - 2x + 1$
   (c) $f(x) = 1 - t - t^4$.
2. Find D^2f if
   (a) $f(x) = 127x - 43x^3$
   (b) $f(x) = (y^3 - 10y)/100$
   (c) $f(x) = \pi(ax^2 - \frac{1}{3}x^3)$
   (d) $f(a) = \pi(ax^2 - \frac{1}{3}x^3)$.
3. Use Theorem 4.5 to find Df if
   (a) $f(x) = (x^2 + 2)(x^3 + 3)$
   (b) $f(x) = (1 - x)(x^2 - 3x + 1)$
   (c) $f(x) = (1 + x^2)^2$
   (d) $f(x) = (1 + x^2)^3$.
4. Express each of the following functions as a composition and apply the chain rule to find Df.
   (a) $f(x) = (x^2 - 2)^3$
   (b) $f(x) = (1 + x^2)^2$
   (c) $f(x) = (1 + x^2)^3$
   (d) $f(x) = (3x^2 - 4)^2 + 5(3x^2 - 4) + 8$

[9] As for the function k, this should also be ${}_x\psi$.

Lemma to Theorem 4.7: Let r be the function

$$r\colon r(x) = \frac{1}{x}, \quad x \neq 0.$$

Then Dr is

$$Dr\colon Dr(x) = -\frac{1}{x^2}, \quad x \neq 0.$$

Proof:

$$\begin{aligned} {}_r\sigma_x(h) &= \frac{1}{h}\left[\frac{1}{x+h} - \frac{1}{x}\right] \\ &= \frac{-1}{x(x+h)} \\ Dr(x) &= \lim_0 {}_r\sigma_x = -\frac{1}{x^2}, \quad x \neq 0. \ \blacksquare \end{aligned}$$

THEOREM 4.7 If f is a differentiable function, then $1/f$ is differentiable, and $D(1/f) = -(Df/f^2)$.

Proof: With the function r of the lemma, write $1/f = r \circ f$. Then

$$\begin{aligned} D\frac{1}{f} &= D(r \circ f) = ((Dr) \circ f)Df \\ &= \left(-\frac{1}{f^2}\right) Df \end{aligned}$$

by Theorem 4.6. ■

Corollary to Theorem 4.7: If f and g are differentiable functions, then f/g is differentiable and $D(f/g) = (1/g^2)(gDf - fDg)$.

Proof: Apply Theorems 4.5 and 4.7 to write

$$\begin{aligned} D\left(\frac{f}{g}\right) &= D\left(f\left(\frac{1}{g}\right)\right) = fD\left(\frac{1}{g}\right) + \left(\frac{1}{g}\right) Df \\ &= -f\left(\frac{Dg}{g^2}\right) + \left(\frac{1}{g}\right) Df = \frac{1}{g^2}(gDf - fDg). \ \blacksquare \end{aligned}$$

THEOREM 4.8 If f is a differentiable function with the inverse function f^{-1}, then f^{-1} is differentiable at each number $x \in \operatorname{dom} f^{-1}$ at which $(Df) \circ f^{-1}(x) \neq 0$, and $Df^{-1} = 1/(Df) \circ f^{-1}$.

Proof: Since $f \circ f^{-1} = I$ for some identity function I, Theorem 4.6 gives

$$((Df) \circ f^{-1})Df^{-1} = D(f \circ f^{-1}) = DI,$$

and hence

$$Df^{-1} = \frac{1}{(Df) \circ f^{-1}}. \blacksquare$$

4.2 PROBLEMS (*continued*)

5. Prove by induction that if $f(x) = x^n$, $n = 1, 2, \ldots$, then $Df(x) = nx^{n-1}$. (See Section 4.3.) Then write at sight a formula for Df if
   (a) $f(x) = x^{10} - x^{20}$
   (b) $f(x) = 9x^{10} - 10x^9 + x^{90}$
   (c) $f(x) = 2(x + x^2) - 3(x^3 - x^4)$
   (d) $f(x) = 100x^{100}$.
6. Use Problem 5 and $x^{-n} = 1/x^n$, $x \neq 0$, $n = 1, 2, 3, \ldots$, to prove that if $f(x) = x^{-n}$, $x \neq 0$, then $Df(x) = (-n)x^{-n+1}$. Then write at sight a formula for Df if
   (a) $f(x) = x^{-4}, \quad x \neq 0$
   (b) $f(x) = 1/x^2, \quad x \neq 0$
   (c) $f(x) = x^{-2} + x^2, \quad x = 0$
   (d) $f(x) = 4/x^3 - 3/x^2 + 2/x + 1, \quad x \neq 0$.
7. Express each function f as a composition and find Df by the chain rule. You specify the domains.
   (a) $f(x) = 1/(1 - x^2)$
   (b) $f(x) = 1/(1 - x^2)^3$
   (c) $f(x) = (x^3 - 2x)^{-4}$
   (d) $f(x) = \left(\frac{1}{x} + \frac{2}{x^2}\right)^3$
   (e) $f(x) = ((5x^4 + 3)^{-2} + 1)^8$
8. Use the corollary to Theorem 4.7 to find Df (you supply the domains) if
   (a) $f(x) = x/(x^3 + 1)$
   (b) $f(x) = \dfrac{x^4 - 3x^2 + 2}{x^3 - 7}$
   (c) $f(x) = \dfrac{\frac{1}{2x} + \frac{1}{3x^2}}{\frac{1}{4x^3} + \frac{1}{5x^4}}$
   (d) $f(x) = \dfrac{x^3 + x^2 + 1}{x^3 - x^2 - 1}$.
9. Define $y = \sqrt{x}$, $x \geq 0$, to be the nonnegative solution of $y^2 = x$ for y. Thus, if $f(y) = y^2$, $y \geq 0$, then $g(x) = \sqrt{x}$, $x \geq 0$ is the inverse of f; that is, $g = f^{-1}$. Apply Theorem 4.8 to deduce that if $x > 0$, $Dg(x) = 1/(2\sqrt{x})$. Then find Df for each of the following.

(a) $f(x) = \sqrt{x+1}$
(b) $f(x) = 3x^{1/2}$
(c) $f(x) = \sqrt{1-2x}$
(d) $f(x) = 1/\sqrt{x}$
(e) $f(x) = (\sqrt{x})^3$

10. Define $y = \sqrt[q]{x}$, $x \geqslant 0$, q a positive integer, to be the nonnegative solution of $y^q = x$. (Denote $\sqrt[q]{x}$ by $x^{1/q}$.) Deduce that if $g(x) = x^{1/q}$, then $Dg(x) = (1/q)x^{(1/q)-1}$. Then find Df for each of the following.
(a) $f(x) = \sqrt[3]{x}$
(b) $f(x) = \sqrt[4]{x}$
(c) $f(x) = \sqrt[10]{x}$
(d) $f(x) = x^{1/5}$
(e) $f(x) = x^{10} + x^{1/10}$
(f) $f(x) = \dfrac{1}{x^3} + \dfrac{1}{x^{1/10}}$

11. View the function $f(x) = x^{p/q} = (x^{1/q})^p$, p,q positive integers, as a composition. Prove that $Df(x) = p/qx^{(p/q)-1}$. (You decide the proper domain.) Then find Df for each of the following.
(a) $f(x) = x^{4/3}$
(b) $f(x) = x^{-3/2}$
(c) $f(x) = (1-x^2)/\sqrt{x}$
(d) $f(x) = 7x^{5/2} + 5x^{3/2}$

12. Prove that if $f(x) = x^r$, on some interval, where r is a rational number, then $Df(x) = rx^{r-1}$ on some domain. (You specify it.) Then find Df if
(a) $f(x) = (x^{1/3} + x^{2/3})(x^{1/4} - x^{1/5})$
(b) $f(x) = (\sqrt{x}+1)(\sqrt[3]{x}-1)$
(c) $f(x) = (x^{100} + x^{50} + x)\left(\dfrac{1}{x} + \dfrac{2}{x^2}\right)$
(d) $f(x) = x^{-4/3} + 5x^{7-3} + 7x^{-5/3}$
(e) $f(x) = \dfrac{(1-\sqrt{x})(1+\sqrt{x})}{1+x^2}$
(f) $f(x) = \dfrac{1-\sqrt{x}}{(1-x^2)(1+\sqrt{x})}$
(g) $f(x) = x^{1/5} - x^{-1/5}$.

The theorems we have proved about differentiable functions are sufficient for doing ordinary differential calculus; they supply the tools. Less and less often are we explicitly making use of the definition of a limit or of a derivative; rather we rely on previous deductions from the real number system axioms and on a precise understanding of the objects dealt with, this understanding being a consequence of clear, unambiguous definitions.

We presume that we need not repeatedly call your attention to the reasons why statements made about equations and inequalities are true. Very likely the role of the least upper bound axiom is less familiar to you, and we beg your indulgence when your attention is called to it. This axiom is basic to analysis (calculus), and in the next chapter you should watch for places where its assumption is crucial.

The next section of this chapter applies the theory of differentiation to the study of special functions; it should be relaxing and fun.

4.3 SOME USES OF THE DERIVATIVE FUNCTION

4.3.1 Polynomial Functions

Consider $T_n(x) = cx^n$, c real, and n a nonnegative integer. If $n = 1$, $T_1(x) = cx$ and $DT_1(x) = c = ncx^{n-1}$ (Theorems 4.2 and 4.4). If $T_n(x) = cx^n$ and $DT_n(x) = ncx^{n-1}$, then[10]

$$\begin{aligned} DT_{n+1}(x) &= D(IT_n)(x) = (IDT_n + T_nDI)(x) \\ &= x(ncx^{n-1}) + cx^n(1) = (n+1)\ cx^n \end{aligned}$$

(Theorem 4.5). Thus, by induction on n we have proved that

$$DT_n(x) = ncx^{n-1}, \quad n = 0,1,2, \ldots,$$

where the $n = 0$ case is supplied by Theorem 4.1.

If P is the polynomial function given by

$$P(x) = \sum_{j=0}^{n} c_j x^j,$$

Theorem 4.3 yields

$$DP(x) = \sum_{j=1}^{n} jc_j x^{j-1},$$

another polynomial function.

It is customary to use an obvious ellipsis in the notation here. $DP(x)$ means, of course, $(DP)(x)$, but if the notation is read $D[P(x)]$, we can get along without using a function symbol whenever we know the correct result ahead of time. To illustrate this, consider $P(x) = 3x^4 - 2x$. We now know that $(DP)(x) = 12x^3 - 2$. This is usually stated and done by simply writing

$$D(3x^4 - 2x) = 12x^3 - 2$$

in which the number symbol, $3x^4 - 2x$ is used in place of the name P of the function it describes.

[10] I is an identity function.

4.3.2 Rational Functions

There is no simple formula for the derivative of the rational function R where

$$R(x) = \frac{P(x)}{Q(x)}, \quad P,Q \text{ polynomial functions}, \quad Q(x) \neq 0.$$

The corollary to Theorem 4.7 tells us that

$$DR(x) = \frac{1}{Q^2}\,[Q(DP) - P(DQ)](x), \quad Q(x) \neq 0,$$

and each such problem is best done by itself. For example,

$$\begin{aligned} D\left(\frac{x^2 - 4x}{x^4 - 3}\right) &= \frac{1}{(x^4-3)^2}\,[(x^4-3)(2x-4) - (x^2-4x)(4x^3)] \\ &= \frac{1}{(x^4-3)^2}\,[-2x^5 + 12x^4 - 6x + 12], \quad x \neq \pm\sqrt[4]{3}. \end{aligned}$$

In some cases the work can be shortened by another theorem. Take the case $D(1/x^3)$ and apply Theorem 4.7 on the domain of the function

$$D\left(\frac{1}{x^3}\right) = -\frac{3x^2}{x^6} = -\frac{3}{x^4}, \quad x \neq 0.$$

In one of the problems you have already found an even better way to look at this one.

4.3.3 Composite Functions

Theorem 4.6 greatly simplifies the work involved in computing the derivative of a function whose formula is complicated. Consider

$$f(x) = (x^3 - 5)^8$$

which is a 24th-degree polynomial function in disguise. The binomial can be expanded and the derivative found by means of the general formula for the derivative of a polynomial function. Or f can be reconstructed as the composition $g \circ h$, where

$$g\colon g(y) = y^8, \qquad h\colon h(x) = x^3 - 5$$

so that

$$Df(x) = D(g \circ h)(x) = ((Dg) \circ h)Dh(x) = Dg(h(x))Dh(x),$$

where $Dg(y) = 8y^7$, $Dh(x) = 3x^2$, and so

$$Df(x) = 8(x^3 - 5)^7\,(3x^2) = 24x^2\,(x^3 - 5)^7.$$

We will prove later on that $D \sin = \cos$, $D \cos = -\sin$, and $D \log = r$, where $r(x) = 1/x$, $x > 0$. With this information and Theorem 4.6, such functions as the following have derivatives which can be found, really trivially.

$$D(\sin \circ r)(x) = D \sin \frac{1}{x} = \left(\cos \frac{1}{x}\right)\left(-\frac{1}{x^2}\right) = -\frac{1}{x^2} \cos \frac{1}{x}, \quad x \neq 0$$

$$D \cos(x^2 + 2x) = -[\sin(x^2 + 2x)](2x + 2) = -(2x + 2) \sin(x^2 + 2x)$$

$$D(\log \circ \cos)(x) = D(\log \cos x) = \frac{1}{\cos x} (-\sin x) = -\tan x, \quad \cos x > 0.$$

The notation we have used in these illustrations is not uniform but is chosen to emphasize the application of Theorem 4.6 without unnecessary proliferation of symbols.

4.3 PROBLEMS

1. Find the derivative of each of the following functions (you specify each domain).
   (a) $h(x) = (2 - 3x^2)^3$
   (b) $h(x) = (3x^2 - 2x + 1)^5$
   (c) $\phi(x) = (x^3 + x^2 + x + 1)^{17}$
   (d) $\phi(y) = (y^{10} - y^{20})^{30}$
   (e) $h(t) = t^{3/2} + (1/t^{3/2})^3$
   (f) $h(x) = (1 - x - x^4)^{-3}$
   (g) $h(y) = \sqrt{y^2 + 1}$
   (h) $f(x) = \left(\dfrac{x}{1 + x}\right)^{3/2}$
   (i) $f(t) = \left(\dfrac{t + 1}{t - 1}\right)^{1/3}$
   (j) $f(x) = (x^2 + 1)^3 (x^3 - 1)^2$
   (k) $\phi(t) = \sqrt{t^3 - 1} \cdot \sqrt[3]{t^2 + 1}$
   (l) $g(z) = \dfrac{\sqrt{1 - z^{10}}}{(10 + z^2)^5}$
   (m) $f(t) = \dfrac{\sqrt{t + (1/t)}\ (t^2 + (1/t^2))^3}{\sqrt[3]{t^3 - 3}\ (t^5 + 5t^4)^6}$
   (n) $\phi(x) = \sqrt{1 + \sqrt{1 - x}}$
   (o) $f(x) = (1 + (2 + x^2)^3)^4$
   (p) $g(t) = (x + \sqrt{1 - x^2})^{2/3}$
   (q) $h(s) = \left(\sqrt{4 + s^2} + \dfrac{1}{\sqrt{1 - s^3}}\right)^{1/4}$
2. Find the second derivative of each of the following functions (you specify each domain).
   (a) $h(x) = \sqrt{4 - x}$

(b) $h(y) = \sqrt{4 - y^2}$
(c) $\phi(t) = (4 - t^2)^{-1/2}$
(d) $\phi(z) = (z - z^2)^{-2}$
(e) $f(z) = (1/(1 + z))^4$

3. Let P be a polynomial function. Prove that each of D^nP is the 0 function if n is greater than the degree of P.
4. In many of the problems above the functions in question could be written as $f \circ g$, where f has the special form $f(x) = x^n$, n some rational number. Prove this theorem:

   If g is a function and n a rational number, then

$$D(g^n) = ng^{n-1}Dg.$$

5. Find the derivative of each of the following functions.
   (a) $\phi(x) = \sin(x^2 + 1)$
   (b) $f(t) = (\sin t)^2$
   (c) $h(y) = \sin\sqrt{y}$
   (d) $\phi(z) = \sqrt{\sin z}$
   (e) $f(x) = x^2 \sin x^2$
   (f) $g(x) = \dfrac{\sin(1/x)}{x}$
   (g) $f(s) = \dfrac{\sin\sqrt{1 + s}}{\sin(1 - s^2)}$
6. Suppose that g is a function such that $Dg = g$.
   (a) Find Df if $f(x) = 1/g(x^{-1})$.
   (b) Find a function f such that $Df(x) = 2xg(x^2)$.
7. Find the derivative of each of the following functions (you specify the domain).
   (a) $f(x) = \log x^2$
   (b) $f(x) = \log^2 x$
   (c) $f(x) = \log|x|$
   (d) $f(x) = \log(1/x)$
   (e) $f(x) = 1/(\log x)$
8. Find a function f such that
   (a) $Df(x) = 3(x + 1)^2$
   (b) $Df(x) = 6(2x - 1)^2$
   (c) $Df(x) = \sqrt{x + 1}, \quad x \geqslant -1$
   (d) $Df(x) = 1/(2\sqrt{x + 1}), \quad x > -1$
   (e) $Df(x) = 1/\sqrt{2x + 1}, \quad x > -\frac{1}{2}$
   (f) $Df(x) = x/\sqrt{x^2 + 1}$.
9. Find D^2h and D^3h if $h = f \circ g$.
10. Find $D^2\phi$ if $\phi = f \circ (g \circ h)$.
11. Assume there is a function log such that

$$D \log = r, \quad r(x) = \frac{1}{x}, \quad x > 0.$$

(a) Prove that log is monotone increasing.
Hint: Consider ${}_{\log}\sigma_x$.
(b) Prove that $\log^{-1}$ exists. Call this inverse function exp with some domain.
(c) Prove that $D \exp = \exp$.

4.3.4 Implicit Functions

Consider the function f given by

$$f(y) = y^2, \quad y > 0.$$

By Theorem 3.17, f has an inverse f^{-1} with the conventional representation

$$f^{-1}(x) = \sqrt{x}, \quad x > 0,$$

and by Theorem 4.8,

$$Df^{-1}(x) = \frac{1}{(Df) \circ f^{-1}(x)} = \frac{1}{2\sqrt{x}}, \quad x > 0.$$

An alternative way to view this problem is to consider the equation

(A) $$x = y^2, \quad \text{or equivalently,} \quad y^2 - x = 0.$$

The set of all ordered pairs of numbers (x,y) which satisfy Equation (A),

(B) $$\{(x,y) \mid y^2 - x = 0\}$$

or the similar set with the order reversed,

(C) $$\{(y,x) \mid y^2 - x = 0\}$$

may (neither one, just one, or both) be the graph of a function. If one of these sets is the graph of a function, that function is said to be given "implicitly" by Equation (A).

In this example set (B) is not the graph of a function since both (1,1) and (1,–1) belong to it. But set (C) is the graph of the function g, given by

$$g(y) = y^2$$

and Equation (A) can be written

(D) $$y^2 - g(y) = 0, \quad \text{for some function } g \text{ and all real } y.$$

Furthermore, (D) asserts that there is a function h, which is a constant function, for which

$$h(y) = 0, \quad \text{all real } y.$$

Therefore,

$$0 = Dh(y) = 2y - Dg(y), \quad \text{that is, } Dg(y) = 2y$$

and we have computed the derivative of g from the implicit relation (A) that defined it.

Let us redefine sets (B) and (C) as follows.

(B′) $\{(x,y) \mid y^2 - x = 0, \quad y > 0\}$ (Therefore, $x > 0$.)
(C′) $\{(y,x) \mid y^2 - x = 0, \quad y > 0\}$ (Therefore, $x > 0$.)

Then both (B′) and (C′) are graphs of functions which are clearly inverse to each other. Let these functions be f and g respectively, so that

$$f^2(x) - x = 0, \quad x > 0$$

and

$$y^2 - g(y) = 0, \quad y > 0$$

and we have before us two constant functions h and j, where

$$h(x) = f^2(x) - x = 0, \quad x > 0$$

and

$$j(y) = y^2 - g(y) = 0, \quad y > 0.$$

Thus the derivatives of both h and j are constant functions, and by the chain rule

$$\begin{aligned} 2f(x)Df(x) - 1 &= 0, \quad x > 0 \\ 2y - Dg(y) &= 0, \quad y > 0. \end{aligned}$$

From the first of these

$$Df(x) = \frac{1}{2f(x)} = \frac{1}{2\sqrt{x}}, \quad x > 0$$

and from the second,

$$Dg(y) = 2y, \quad y > 0.$$

More generally, consider $F(x,y)$, where this symbol stands for some (simple or complicated) formula, and suppose that

$$F(x,y) = 0 \quad \text{for some pairs of numbers } x,y.$$

If the set

$$\{(x,y)\,|\,F(x,y)=0\}$$

is the graph of a differentiable function f, then

$$h(x)=F(x,f(x))=0$$

and h has a derivative Dh, which can lead to a formula for Df.

The existence of the function f is the crucial matter, and in practice this is often assumed in order to "get ahead."[11] As a rather transparent example of what this can lead to, consider

$$x^2+y^2+4=0.$$

Assuming a function f, so that

$$x^2+f^2(x)+4=0, \quad a \leqslant x \leqslant b,$$

say, we find the derivative of f by

$$2x+2f(x)\;Df(x)=0$$

and so

$$Df(x)=-\frac{x}{f(x)}, \quad a \leqslant x \leqslant b, \quad f(x)\neq 0,$$

where neither f nor Df exists.

4.3 PROBLEMS (*continued*)

12. Assume there is a function f so that $y=f(x)$ on some domain. Find a formula for Df.
    (a) $x^2+y^2-1=0$
    (b) $4x^2+9y^2-36=0$
    (c) $\sqrt{x}+\sqrt{y}=5$
    (d) $x^{2/3}+y^{2/3}=1$
    (e) $x^3+x^2y^2+y^3=1$
    (f) $x+2\sqrt{xy}+x/y=x+y$
    (g) $\sqrt{1+xy}=(1+(x/y))^3$
    (h) $\sqrt{x^2y^2-1}+x^3y^3\;((1+x)/y)^2=0$
13. Proceed as in Problem 12 for the following.
    (a) $\cos(x+y)+\sin(x+y)=0$
    (b) $\cos xy+\sin xy=0$
    (c) $\log(x+y)+\exp(x+y)=0$
    (d) $\log xy+\exp xy=0$

[11] Look for an "implicit function" theorem in a more advanced course to justify such an assumption.

4.3.5 Tangent Lines

If f is a differentiable function, then at a number $x \in \operatorname{dom} f$ the slope function ${}_f\sigma_x$ measures the slope of a line through two points on the graph of f, as we have seen. A value of ${}_f\sigma_x$ near 0 corresponds to such a line through two points on the graph which are close together. This motivates the next definition.

DEFINITION 4.5 The line through the point $Q(x,f(x))$ with slope $Df(x)$ is *tangent* to the graph of f at Q. It is called the *tangent line* at Q.

If $P(X,Y)$ is a point on the tangent line at Q, then it is easy to see that an equation of this tangent line is

$$\text{(A)} \qquad Df(x)\ X - Y + f(x) - xDf(x) = 0.$$

As an example, consider the circle $x^2 + y^2 = 1$. A differentiable function implicit in this relation is f,

$$f(x) = \sqrt{1 - x^2}, \quad -1 < x < 1,$$

where $x^2 + f^2(x) = 1$. Then

$$2x + 2f(x)\ Df(x) = 0$$

and

$$Df(x) = \frac{-x}{f(x)}$$

so that from Equation (A) the tangent line to the top semicircle at a point $Q(x,f(x))$, $-1 < x < 1$ has the equation

$$xX + f(x)Y - 1 = 0.$$

See Figure 4.5.

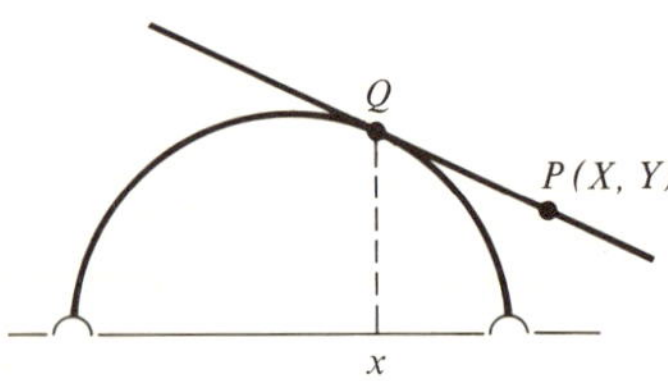

Figure 4.5

In a more easily remembered form this tangent line is given by

$$xX + yY = 1, \quad Q(x,y) \text{ is on the circle,} \quad P(X,Y) \text{ is on the tangent line.}$$

It will be of use soon to observe that a tangent line at Q is parallel to the x-axis (horizontal) if, and only if, $Df(x) = 0$, and its equation is $Y = f(x)$ or, simply, $Y = y$. Also, Definition 4.5, which refers only to differentiable functions, never yields a tangent line parallel to the y-axis (vertical).

When we study geometric curves, a parametric representation is often useful. Suppose that a curve C is given by

$$x = \phi(t), \quad y = \psi(t), \quad a \leqslant t \leqslant b.$$

This means that the curve is the set of points $\{(\phi(t), \psi(t)) | a \leqslant t \leqslant b\}$. Suppose that ϕ, ψ are differentiable functions and one of them, say ϕ, has an inverse ϕ^{-1}. Then ϕ^{-1} is differentiable, subject to the condition in Theorem 4.8, and

$$y = \psi \circ \phi^{-1}(x) = f(x)$$

so that Equation (A) will give the tangent line to C at the point $Q(x,f(x))$. Given t, both x and $f(x)$ are known. So this point Q on C can be expressed as $Q(t)$. To apply Equation (A) we compute $Df(x)$ by the chain rule and Theorem 4.8.

$$\begin{aligned} Df(x) &= ((D\psi) \circ \phi^{-1}) D\phi^{-1}(x) \\ &= ((D\psi) \circ \phi^{-1}) \frac{1}{(D\phi) \circ \phi^{-1}} (x) \end{aligned}$$

so that for each pair of numbers t,x satisfying $x = \phi(t)$,

$$Df(x) = \frac{D\psi(t)}{D\phi(t)}, \quad D\phi(t) \neq 0,$$

and Equation (A) implies that

$$\text{(B)} \qquad D\psi(t)X - D\phi(t)Y + [\psi D\phi - \phi D\psi](t) = 0$$

is the equation of the tangent line to C at $Q(t)$.

If it should happen that $D\phi(t) = 0$ but $D\psi(t) \neq 0$, we call the line (B) a *vertical tangent* to C at $Q(t)$. In case both these derivatives are 0 at t, then all is lost.

In the example of the semicircle suppose that we include the end points:

$$C\colon x = \cos t, \quad y = \sin t, \quad 0 \leqslant t \leqslant \pi.$$

Here, $D\phi(t) = -\sin t$, $D\psi(t) = \cos t$, and Equation (B) gives

$$X \cos t + Y \sin t = 1, \quad 0 \leqslant t \leqslant \pi.$$

At $t = 0$ and $t = \pi$ the vertical tangent lines $X = 1$ and $X = -1$ appear. The derivatives never vanish together.

4.3 PROBLEMS (*continued*)

14. Find the tangent lines to the graphs of the following functions at the points indicated.
 (a) $f(x) = 2x^2 - 6x + 1, \quad x = 1$
 (b) $f(x) = 3 - 2x, \quad x = -2$
 Answer: $2x + y - 3 = 0$
 (c) $f(x) = x^3 - 3x^2 + 1, \quad x = 2$
 (d) $g(x) = x^4 + 3x^2 - 6, \quad x = 0$
 (e) $f(y) = \sqrt{y}, \quad y = 9$
 Answer: $x - 6y + 9 = 0$
 (f) $g(t) = t(1 + |t|), \quad t = 0$
 (g) $h(t) = t^{15} - t^7 + 1, \quad t = 1$
 (h) $\phi(z) = 2z^2 - (2/z^2), \quad z = -1$
 (i) $g(y) = y/(1 + y), \quad y = 5$
 Answer: $x - 36y + 25 = 0$
 (j) $h(w) = \sqrt{w}/(1 + \sqrt{w}), \quad w = 4$
 (k) $\phi(x) = (1 - x^2)(1 + x + x^3), \quad x = -2$
 Answer: $75x + y + 123 = 0$
 (l) $f(z) = (1 - z^3)^2, \quad z = \frac{1}{2}$
 (m) $g(z) = (z^{1/2} + z^{1/4})(z^{3/4} + z^{4/3}), \quad z = 1$
 (n) $g(t) = (t - (1/t^2))(t^2 + (2/t)), \quad t = -4$
 (o) $h(t) = (t^2/(1 + t^3)), \quad t = 5$
 (p) $h(y) = (y^2 + 1)/(y^2 - 4), \quad y = -1$
 (q) $\phi(y) = (1 + \sqrt{y})/(y - \sqrt{y}), \quad y = 16$
15. Find the equations of the normals (lines perpendicular to the tangents at the point of tangency) in Problem 1.
16. For each of the following functions find where the tangent line is horizontal.
 (a) $f(x) = x^3 - 3x^2 + 1$
 Answer: $(0,1)$ and $(2,-3)$
 (b) $g(x) = x^4 - 8x^2 + 5$
 (c) $h(t) = t^{4/3}$
 (d) $f(y) = y/(4 - y)$
 (e) $g(y) = y^2/(4 - y)$
 Answer: $(0,0)$ and $(8,-16)$
 (f) $h(z) = (1 - \sqrt{z})/(1 + \sqrt{z})$
17. Find the equations of the tangent lines at the indicated points to the graphs of functions defined by the following equations.
 (a) $x^2 + xy + 2y^2 = 28, \quad (2,3)$
 Answer: $x + 2y - 8 = 0$
 (b) $x^3 + 3xy^2 + y^3 = 1, \quad (2,-1)$
 (c) $\sqrt{4x} - \sqrt{9y} + 5 = 0, \quad (4,9)$

Answer: $x - y + 5 = 0$

(d) $x\sqrt{xy} + 2y^2 - 3 = 0, \quad (1,1)$

(e) $\sqrt{3 + x^2y^2} - 2x^2/y^2 = 0, \quad (1,1)$

(f) $y^2 = 4px, \quad (p,-2p)$

18. Find all the points on the graph of $f(x) = x^3 + 2x^2 + 3x + 4$ where the slope of the tangent line is 2.
19. Find all points on the graph of $f(x) = x^3 - 3x^2 + 1$ where the tangent is parallel to the line $9x - y + 5 = 0$.
20. Find all points on the graph of $f(x) = x^2/(1 + x)$ where the tangent is perpendicular to the line $x - 8y + 2 = 0$.
21. Show that the slope of the graph of $f(x) = ax^3 + bx + c$ is never 0 if $ab > 0$.
22. The tangent line to the graph of $f(x) = x^3 - 3x^2 + 3x$ at the point (2,2) meets the graph at another point. Find this point.
23. The graph of $f(x) = x^2 + ax + b$ is tangent to the line $y = 2$ at the point (4,2). Find a and b.
24. For which values of a, b, and c is it true that the graphs of $f(x) = x^2 + ax + b$ and $g(x) = x^2 + cx$ have the same tangent line at the point (2,2)?
25. Over which points of the x-axis are the tangents to the graphs of $f(x) = x^3 + 8$ and $g(x) = 3x^2 - 3x + 9$ parallel?
26. The graphs of $f(x) = x^3 + x$ and $g(x) = x^2$ meet at the origin. What is the angle between the tangents to the graphs of these functions at this point?

    Remark: The angle between the tangents to two curves at a point of intersection of the curves is called the *angle of intersection* of the curves.
27. Find the points of intersection of the following pairs of curves and, at each such point, find the tangent of the angle of intersection.

    (a) $f(x) = x^3 + 3x + 1, \quad g(x) = 3x$

    Answer: $(-1,-3)$; 3/19.

    (b) $h(x) = x^2, \quad \phi(x) = \sqrt{x}$

    (c) $f(t) = 4 - t^2, \quad h(t) = 2 - t^2/2$

    (d) $g(z) = z^2/4 - 1, \quad h(y) = 4 - y^2/16$
28. Find the equations of the tangent lines to the graph of $f(x) = x^2$ which pass through the point $(0,-4)$.

    Hint: It is necessary to find the points of tangency. If (a,b) is such a point, we have one equation connecting a and b from the fact that the point is on the curve. Another equation connecting these two unknowns comes from the fact that the slope of the tangent line is $Df(a)$ and is also $(b + 4)/a$.

    Answer: $4x - y - 4 = 0$ and $4x + y + 4 = 0$
29. Find the equations of the tangent lines to the graph of $f(x) = x^2$ which pass through the point $(1,-3)$.

30. Find the equation of the normal line to the graph of $f(x) = x^2$ which passes through the point (3,0).
31. Show that the two lines which pass through the point $(\frac{3}{2},0)$ and are tangent to the graph of the function $g(x) = (x^2 + 4)/4$ are perpendicular.
32. Find the slopes of the three lines which pass through the point (4,8) and are normal to the graph of the function $h(x) = (x^2 - 4x)/8$.
33. If the set of points (x,y) for which

$$Ax^2 + Bxy + Cy^2 + Dx + Ey + F = 0$$

is not empty, then that set is called a conic. Prove that the tangent line to the conic at a point $Q(x,y)$ on it is

$$AxX + (\tfrac{1}{2})B(yX + xY) + CyY + (\tfrac{1}{2})D(x + X) + (\tfrac{1}{2})E(y + Y) + F = 0,$$

where $P(X,Y)$ is a point on the tangent line.
34. Find several tangent lines to
    (a) the circle $(x - 1)^2 + y^2 = 1$
    (b) the circle $(x - 1)^2 + y^2 = 4$
    (c) the ellipse $\frac{1}{4}(x - 2)^2 + \frac{1}{9}(y + 1)^2 = 1$
    (d) the parabola $16x^2 + 24xy + 9y^2 + 15x - 20y = 0$.
35. Let ϕ and ψ be differentiable functions on a common domain, and $x = \phi(t)$, $y = \psi(t)$, $f = \psi \circ \phi^{-1}$, $g = \phi \circ \psi^{-1}$. For each of the following examples describe a common domain for ϕ and ψ, write formulas for f, g, Df, and Dg. Write formulas for tangent lines to the graphs of f and g.
    (a) $x = t, \quad y = 1/t$
    (b) $x = t, \quad y = t^2$
    (c) $x = \cos t, \quad y = \sin t$
    (d) $x = \cos t, \quad y = \cos 2t$

4.3.6 Differential of a Function

Consider two functions f and g for which $\lim_a f = \lim_a g = 0$. Suppose that for each h near the number a, $f(h)$ is closer to 0 than $g(h)$ is. Loosely speaking, if h is taken successively closer to a and if $f(h)$ and $g(h)$ are racing toward 0, $f(h)$ would appear to stay ahead. We now formulate a measure to tell how $f(h)$ leads $g(h)$ as they race toward 0.

For example, suppose that the two functions f and g are

$$f(h) = \tfrac{1}{2}(h - 1), \quad g(h) = (h - 1), \quad \text{near } a = 1.$$

Clearly, $\lim_1 f = \lim_1 g = 0$, and $|f(h)| < |g(h)|$, h near 1. Let us look at the limit of their ratio;

$$\lim_{1} \frac{f}{g} = \lim_{h \to 1} \frac{1}{2} = \frac{1}{2}.$$

As another example, consider

$$f(h) = (h-1)^2, \quad g(h) = h - 1, \quad \text{near } a = 1.$$

Again, $\lim_{1} f = \lim_{1} g = 0$, $|f(h)| = |h-1|^2 < |h-1| = |g(h)|$ for $|h-1| < 1$, and the limit of their ratio is

$$\lim_{1} \frac{f}{g} = \lim_{h \to 1} (h-1) = 0,$$

0 this time. In view of these examples we formulate the following definition.

DEFINITION 4.6 If $\lim_{a} f = \lim_{a} g = 0$ and $\lim_{a}(f/g) = 0$, then f "goes to 0 faster" than g.[12]

Let f be a differentiable function and $Q(x, f(x))$ be a point on its graph. The tangent line to the graph at Q is

$$Y = Df(x)X + f(x) - xDf(x), \quad P(X,Y) \text{ on tangent line.}$$

If a coordinate system is chosen with the origin at Q, then this line is a graph of the linear function ${}_xd$.

$$\text{(C)} \qquad {}_xd(h) = Df(x)h.$$

Assume that $f(x)$ and $Df(x)$ are given and that we desire to compute $f(x+h)$ at a number near x. If $e(h)$ is the error we make by using $f(x) + {}_xd(h)$ instead of $f(x+h)$; that is, by using the tangent line instead of the curve, then $f(x+h) = f(x) + {}_xd(h) + e(h)$. See Figure 4.6. Clearly,

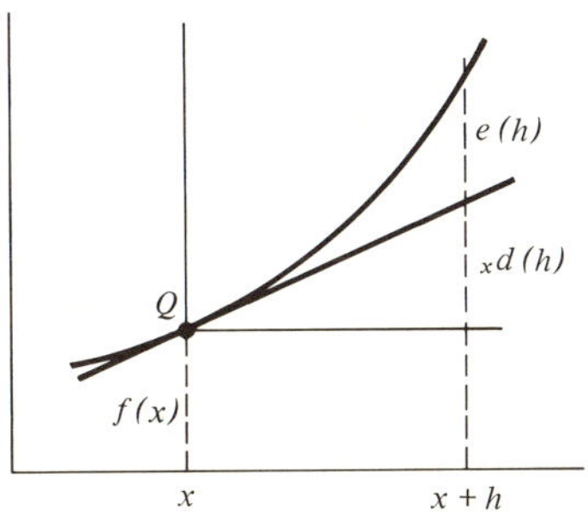

Figure 4.6

[12] This is recorded in some books by $f = 0(g)$.

$$\text{(D)} \qquad \lim_0 e = \lim_{h\to 0}[f(x+h) - f(x) - {}_xd(h)]$$
$$= \lim_{h\to 0} h[{}_f\sigma_x(h) - Df(x)] = 0$$

and if an error $e(h)$ is tolerable, there will be an interval $(x - \delta, x + \delta)$; that is, $|h| < \delta$, within which $f(x) + {}_xd(h)$ will be acceptable for $f(x + h)$.

Try this with $f(x) = x^2$, for $x = 3$. Then

$$f(x) = f(3) = 9, \quad Df(x) = 2x = 6, \quad {}_3d(h) = 6h.$$

Suppose that we can tolerate an error of 0.01 in computing squares of numbers near 3. In line (D)

$$|f(3+h) - f(3) - {}_3d(h)| = h^2 < 0.01$$

in case $|h| < 0.1$. Therefore, $(3 + h)^2 = 9 + 6h + e(h)$, with $|e(h)| < 0.01$ when $|h| < 0.1$.

This process of replacing a differentiable function by a linear function, $f(x + h)$ by $f(x) + {}_xd(h)$, is best served by the particular linear function ${}_xd$. To see why, we ask how fast the error goes to 0 with h. Let $e(h)$ be the error with ${}_xd$ and $\bar{e}(h)$ be the error with some other linear function ${}_x\bar{d}$. Then

$${}_xd(h) = Df(x)h, \quad {}_x\bar{d}(h) = ch, \quad \text{for some } c \neq Df(x)$$
$$e(h) = h[{}_f\sigma_x(h) - Df(x)], \quad \bar{e}(h) = h[{}_f\sigma_x(h) - c]$$

and

$$\lim_0 e = \lim_0 \bar{e} = 0.$$

But

$$\lim_0 \frac{e}{\bar{e}} = \lim_{h\to 0} \frac{{}_f\sigma_x(h) - Df(x)}{{}_f\sigma_x(h) - c} = 0$$

so the error *goes to 0* faster with the ${}_xd$ function. Geometrically, we say that the tangent line is *closer* to the curve than any other line through the point of tangency.

DEFINITION 4.7 The *differential* of a differentiable function f is a function whose domain is a set of ordered pairs of numbers (specified below) and whose rule is

$$df\colon df(x,h) = Df(x)h, \quad x \in \operatorname{dom} f, \quad h \text{ real}.$$

This differential is a function of *two variables*.[13] For our purposes we need only to speak of the differential of f at x, which is our special linear function ${}_xd$ of the *one variable, h.*

Although it is true that $df(x,h) = {}_xd(h)$ for $x \in \text{dom } Df$ and all real h, the two functions df and ${}_xd$ are different creatures.

4.3 PROBLEMS (*continued*)

36. Of each pair of functions, does one of them go to 0 faster than the other?
    (a) $f(x) = x - 1, \quad g(x) = k(x-1), \quad k \neq 0$ at 1
    (b) $f(x) = x + 3, \quad g(x) = (x+3)^2$ at -3
    (c) $f(x) = x - 4, \quad g(x) = (x-4)^{1/3}$ at 4
    (d) $f(x) = (x+7)^4, \quad g(x) = (x+7)^{4.001}$ at -7
    (e) $f(x) = x, \quad g(x) = \sin x$ at 0
    (see Section 3.1.)
    (f) $f(x) = x^{1.001}, \quad g(x) = \sin x$ at 0
37. Assume $D \sin = \cos$. Prove that, given $\epsilon > 0$, there is an interval containing zero in which

$$1 - \epsilon < \frac{\sin h}{h} < 1 + \epsilon.$$

This verifies the statement made on page 65.

38. Assume that

$$D \log x = 1/x.$$

Prove that, given $\epsilon > 0$, there is an interval containing zero in which

$$1 - \epsilon < \frac{\log(1+h)}{h} < 1 + \epsilon.$$

39. We wish to compute the cubes of numbers near 2, with an error less than 0.1. Show that for numbers between 1.9 and 2.1, the formula $8 + 12h$ for $(2+h)^3$ will do this. Test this with $(2.05)^3$. Find $(1.97)^3$ correct to one decimal.
40. We wish to compute reciprocals of numbers near 100, with an error less than 0.00001. Show that for numbers between 99 and 101, the formula $0.0001(100 - h)$ for $1/(100 + h)$ will do this. Test this with 1/99.5 and 1/100.5.

[13] See Appendix III. As it occurs above the word "variable" is used in a conventional sense. A comment on this appears in the first paragraph of Appendix III.

4.3.7 Some Older Notation

If an element in the graph of a function f is the ordered pair of numbers (x,y), then $y = f(x)$. It has long been customary to slur the distinction between the function f and the number y and to let the letter y connote the function itself as well.[14] The symbol dy/dx is used to represent both $Df(x)$ and Df, similarly. Thus we read, "Let dy/dx be the derivative of $y = f(x)$." One way to render this symbol usable in the function context is to view the d in the denominator as the initial letter of "domain." Then, when confronted with dy/dx, we assume a differentiable function f is present, x is to stand for domain elements, and either $dy/dx = Df(x)$ or $dy/dx = Df$, usually the former.

For example, if $dy/dx = ax + b$, some function such as $f(x) = \frac{1}{2}ax^2 + bx + 3$ is present, and it is agreed that $y = f(x)$. If $dy/db = ax + b$, then possibly $g(b) = abx + \frac{1}{2}b^2 + 4$ and $y = g(b)$.

If dy/dx is broken apart as if it were a fraction, then $dy/dx = Df(x)$ leads to $dy = Df(x)dx$. Comparing this with $df(x,h) = Df(x)h$ of our differential discussion, we can interpret the couplet dx to be h, and dy to be $df(x,h)$. The couplet dy is often called the *differential of* y (meaning the differential of f) and is alternatively used for the two variable function df and for the number $df(x,h)$. Symbolically, the identity

$$dy = \frac{dy}{dx}\,dx$$

is very clean, but needs careful interpretation.

For many purposes, when a function f is present, it is useful to let the pair dx,dy stand for any pair of numbers, with $dx \neq 0$, whose ratio $dy/dx = Df(x)$ at some x. See Figure 4.7. This is a kind of all-purpose

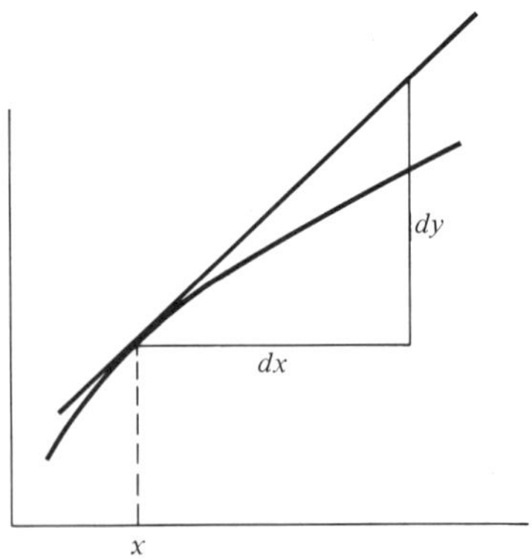

Figure 4.7

[14] You will find even $y = y(x)$ in many calculus books.

pair of symbols; it is no wonder that it has persisted in the literature. In proofs of theorems about derivatives it is usually more trouble to describe just what is meant by dx and dy at the time than it is to use the more well defined function notation. However, the results of such proofs can often be recorded very neatly with them.

EXAMPLE If $u = f(x)$, $v = g(x)$, then the sum, product, and quotient formulas for derivatives are

$$\frac{d(u+v)}{dx} = \frac{du}{dx} + \frac{dv}{dx}$$

$$\frac{d(uv)}{dx} = v\frac{du}{dx} + u\frac{dv}{dx}$$

$$\frac{d\left(\frac{u}{v}\right)}{dx} = \frac{v\frac{du}{dx} - u\frac{dv}{dx}}{v^2}.$$

EXAMPLE If $y = f(x)$, $x = g(t)$, so that $y = f(g(t))$, then the chain rule takes the form

$$\frac{dy}{dt} = \frac{dy}{dx}\frac{dx}{dt}.$$

EXAMPLE If $y = f(x)$, so that $x = f^{-1}(y)$, the rule for finding the derivative of f^{-1} looks like

$$\frac{dx}{dy} = \frac{1}{\frac{dy}{dx}}.$$

In some writing the fraction symbol dy/dx is considered too cumbersome, and a primed notation is used. So if $y = f(x)$, then

$$y' = \frac{dy}{dx} = Df(x).$$

This is handy for implicit differentiation. If $x^2 + y^2 = 1$ is assumed to define a function $y = f(x)$, then $Df(x)$ is quickly computed by

$$2x + 2yy' = 0, \quad y' = \frac{-x}{y}.$$

Suppose that $y = f(x)$, $x = g(t)$, as in the example above. Then you may see

$$\dot{y} = y'\dot{x},$$

where the prime refers to an *x-derivative* and the dot refers to a *t-derivative*. This is just a shorthand record of an application of the chain rule.

4.3 PROBLEMS (*continued*)

Do these problems in the "older notation," and accompany each solution with a description of the functions present and what the solution means.

41. Find dy/dx for each of
    (a) $y = x^2 - 4x + 1$
    (b) $y = \sin(x^2 + 4)$
    (c) $y = (x^3 - 2x)/(x^4 + 1)$
    (d) $y = \log \cos (1/x)$
    (e) $y = \sin^{-1} x$.
42. Set $x = 1$ in each of the parts of Problem 40 and write three values of the pairs (dy, dx).
43. Explain what $dy/dt = dy/du\ du/dv\ dv/dt$ should mean.
44. Explain what

$$\text{if } y = u^n, \text{ then } \frac{dy}{dx} = nu^{n-1}\frac{du}{dx}$$

should mean.
45. Interpret the expression $(x - y^2)dx + (y - x^2)dy = 0$.
46. Find a function $y = f(x)$ so that
    (a) $xdx + ydy = 0$
    (b) $y^2dx + x^2dy = 0$.

4.3.8 Rates

Suppose that a physical particle moves on a line so that its distance (x feet) from the 0 point on the line is related to the time (t seconds) at which it attains the position x by

$$x = 3t - 4.$$

We have agreed previously that the velocity of this particle at time t is given by

$$\frac{dx}{dt} = 3.$$

Over any one second time interval the particle actually moves

$$x(t + 1) - x(t) = [3(t + 1) - 4] - [3t - 4] = 3 \text{ feet},$$

that is, 3 feet in each second. Briefly, the particle moves at a rate of 3 feet per second, and this is the way velocity is described.

Now, suppose that the initial relation were

$$x = 16t^2.$$

Then the velocity would be given by

$$\frac{dx}{dt} = 32t.$$

Over a 1 second time interval the particle moves

$$x(t+1) - x(t) = 16(t+1)^2 - 16t^2 = 32t + 16 \text{ feet},$$

clearly depending on the time at which the one second interval starts. However, at no time does the formula for the velocity give this distance moved in 1 second ($32t = 32t + 16$ has no solution). So what does the velocity formula say?

Consider the time interval h and suppose that the displacement formula describes a differentiable function (which it does in this example) so the displacement over the time interval h is expressible by the differential of this function as

$$x(t+h) - x(t) = dx(t,h) + {}_te(h),$$

where the error ${}_te$ goes to 0 faster than h does. The ratio of the distance moved to the time interval h

$$\frac{1}{h}\,[x(t+h) - x(t)] = \frac{dx}{dt} + \frac{1}{h}\,{}_te(h) = 32t + \frac{1}{h}\,{}_te(h)$$

is then given by the velocity formula, $32t$, with an error that goes to 0 with the time interval h. This is what the velocity formula says about the distance-to-time ratio. The term "feet per second" is a literally accurate description of the velocity function only when it is constant, as when the displacement function is a linear polynomial function. This term is artificially (and universally) used to describe "how fast" a particle is moving when the displacement function is differentiable.

More generally, if a differentiable function f describes the relation between the measures x (unit_1) and y (unit_2) of two quantities

$$y = f(x),$$

then

$$\frac{1}{h}\,[f(x+h) - f(x)] = Df(x) + \frac{1}{h}\,{}_xe(h),$$

where $\lim\limits_{h \to 0} (1/h)\,{}_xe(h) = 0$. The function Df is called the *rate function* for f, and its value $Df(x)$ is the *rate of change of f (or y) with respect to x.* This value is assigned the unit, unit_1 per unit_2, which is written $\text{unit}_1/\text{unit}_2$.

4.3 PROBLEMS (*continued*)

47. A particle moves on a line, its position at any time t being $s(t) = t^3 - 6t^2 + 9t - 4$. Find when the velocity is 0, when it is positive, and when it is negative. The motion of the particle for $0 \leqslant t \leqslant 4$ is represented graphically by Figure 4.8.

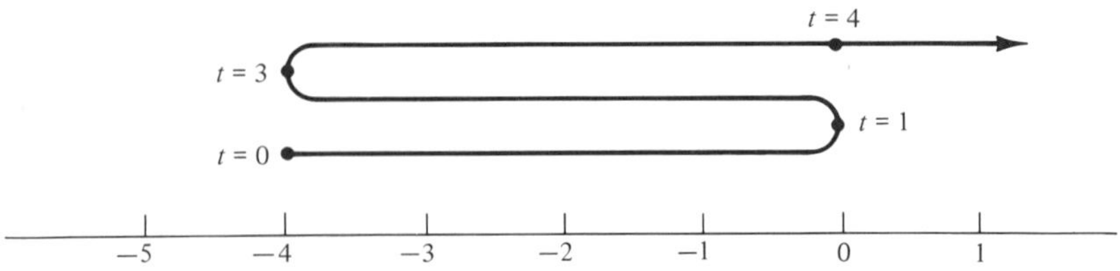

Figure 4.8

48. For each of the following laws of motion, investigate the velocity as in Problem 45 and represent the motion graphically for the time range indicated.
    (a) $s(t) = t^2 + 4t + 3, \quad -4 \leqslant t \leqslant 4$
    (b) $s(t) = t^3 - 2t^2 - 4t + 4, \quad -3 \leqslant t \leqslant 3$
    (c) $s(t) = 64t - 16t^2, \quad 0 \leqslant t \leqslant 4$
    (d) $s(t) = t - 4/\sqrt{t+1}, \quad 0 \leqslant t \leqslant 8$
49. In each case of Problem 46 find when the acceleration is 0.
50. In each case of Problem 46 find $(s(t+h) - s(t))/h$ for the intervals $t = 0$ to $t = 1$, $t = 1$ to $t = 2$, and $t = 0$ to $t = 2$.
51. A cube is expanding in such a way that its edge is changing at a rate of 5 inches/second. When its edge is 4 inches long, find the rate of change of its volume.
52. A metal sphere is heated so that its radius increases at the rate of 1 millimeter per second. How fast is its volume changing when its radius is 3 centimeters?
53. The area of an equilateral triangle is changing at the rate of 1 square inch per second. How fast is its side changing when its area is $100\sqrt{3}$? *Answer:* $\sqrt{3}/30$
54. A point P moves along the graph of $(x^2 + 4)y = 8$ so that its abscissa is changing at a rate of 3 units per second. At what rate is the ordinate changing when $x = 2$?
55. In Problem 53 at what rate is the slope of the curve at P changing when $x = 2$? when $x = 0$?
56. In Problem 53 consider the line PQ, where $Q = (0,4)$.
    (a) At what rate is the distance between P and Q changing when $x = 2$?
    (b) At what rate is the slope of the line PQ changing when $x = 2$?
57. Work Problem 55 (b) if the ordinate, instead of the abscissa, is changing at the rate of -4 units per second.
58. A winch located at the end of a dock 6 feet above the water hauls in a rope at the rate of 2 feet per second. How fast is a boat that is attached to the end of the rope approaching the dock when there are 10 feet of the rope out?

59. Find the rate of change of the area of a circle with respect to
    (a) its radius;
    (b) its diameter;
    (c) its circumference.
60. The radius r and altitude h of a right circular cylinder are changing respectively at the rates of 2 and 3 inches/second.
    (a) At what rate is the volume changing when $r = 4$ and $h = 22$?
    (b) At what rate is the total surface area changing when $r = 4$ and $h = 2$?
61. A particle moves on a line with a constant acceleration of α units per second.
    (a) If it has a velocity of v_0 units per second at time $t = 0$, find the velocity at any time.
    (b) If it is at the origin when $t = 0$, find its position at any time.

Chapter 5 EXTREME VALUES OF A FUNCTION

5.1 INTRODUCTION

We shall now examine the ranges of functions. One simple question we can ask about a particular function is, "Does its range contain a largest number?" "A least number?" Sometimes this is easy to answer.

EXAMPLE 1 $\quad f(x) = 1/(x^2 + 1), \quad x$ real

Since $x^2 + 1 \geqslant 1$, $f(x) \leqslant 1 = f(0)$ for all $x \in \operatorname{dom} f$ and 1 is the largest number in range f. Also, for any $x > 0$,

$$0 < f(x + 1) = \frac{1}{(x^2 + 1) + 2x + 1} < f(x).$$

Hence there is no least number in range f.

EXAMPLE 2 $\quad g(x) = \dfrac{1}{|x|}, \quad x \neq 0$

$\qquad g(0) = 1$

Since $g(\frac{1}{2}) = 2 > 1 = g(0) > \frac{1}{2} = g(2)$, $g(0)$ is neither the greatest nor the least number in range g. For $x > 0$,

$$g\left(\frac{x}{2}\right) = \frac{2}{x} > \frac{1}{x} = g(x) > \frac{1}{2x} = g(2x)$$

and for $x < 0$ these inequalities are simply reversed. So there is no greatest or least number in range g at all.

However, the set

$$\{g(x) \mid |x| \leq \tfrac{1}{2}\} = [2,\infty) \cup \{1\}$$

does have a least number; that is, 1. This number is the least range number of all those for $|x| \leq \frac{1}{2}$.

DEFINITION 5.1 If there is a number $\bar{x}$ in dom f for which

$$f(x) \leq f(\bar{x}), \quad x \in \operatorname{dom} f,$$

then $f(\bar{x})$ is the *absolute maximum* value of f.

DEFINITION 5.2 If there is a number $\bar{x}$ such that for some open interval I, containing $\bar{x}$,

$$f(x) \leq f(\bar{x}), \quad x \in I \cap \operatorname{dom} f,$$

then $f(\bar{x})$ is a *local maximum* value[1] of f at $\bar{x}$.

Absolute and local minima are defined symmetrically. In the examples above, $1 = f(0)$ is an absolute maximum of f, and $1 = g(0)$ is a local minimum of g.

5.1 PROBLEMS

Find all local and absolute extremes of each of the following functions.

1. $f(x) = 2x, \quad 0 \leq x \leq 1$
2. $f(x) = 2x, \quad 0 < x \leq 1$
3. $f(x) = 2x, \quad 0 < x < 1$
4. $$\begin{aligned} f(x) &= 2x \quad &, \quad 0 \leq x < 1 \\ &= \tfrac{1}{2} &, \quad x = 1 \\ &= -x + 2, &\quad 1 < x \leq 2 \end{aligned}$$

[1] This is also called a *relative maximum* value of f—relative to some open interval I, that is. Henceforth we will often omit the word "value" with reference to these extreme values of functions.

5. $f(x) = 2x \quad , \quad 0 < x < 1$
   $\quad\;\; = -x + 2, \quad 1 < x < 2$
6. $f(x) = 2x, \quad 0 < x \leqslant 1$
   $\quad\;\; = 2 \;, \quad 1 < x < \tfrac{3}{2}$
   $\quad\;\; = 1 \;, \quad x = \tfrac{3}{2}$
   $\quad\;\; = 2 \;, \quad \tfrac{3}{2} < x < 2$
7. $f(x) = 1, \quad 0 \leqslant x \leqslant 1, \quad x$ rational
   $\quad\;\; = 0, \quad 0 \leqslant x \leqslant 1, \quad x$ irrational
8. $\operatorname{dom} f = [0,1]$ and
   $f(x) = 0 \quad , \quad x$ irrational
   $\quad\;\; = 1/q, \quad x = p/q, \quad \text{g.c.d}(p,q) = 1$
   $\quad\;\; = 1 \quad , \quad x = 0 \quad , \quad x = 1$

5.2 THE MEAN VALUE THEOREM

There is no general criterion, other than the definitions themselves, that suffices to produce the extreme values of all functions. However, in the class of differentiable functions we do now develop some usable criteria for finding extreme values. Since a differentiable function is necessarily continuous, we start by proving a strong result about continuous functions.

EXTREME VALUE LEMMA If f is a continuous function whose domain is a closed interval, then f has an absolute maximum and an absolute minimum.

Proof: To begin with we will establish that there is a number M such that $|f(x)| \leqslant M$, $x \in \operatorname{dom} f$; that is, f is a bounded function. Let $\operatorname{dom} f = [a,b]$.

Define the set C.

$$C = \{c | f \text{ is bounded in } [a,c]\}.$$

Since $f(a)$ exists, $a \in C$ and $C \neq \varnothing$. Also, $C \subset [a,b]$ and $b = \text{ub } C$ exists. The least upper bound axiom assures the existence of a number $c^* = \text{lub } C$. If we can decide that $c^* = b$, with $c^* \in C$, then we will know that $[a,b] = C$, which means that f is a bounded function.

Since f is continuous at c^*, there is a $\delta > 0$ for which $|x - c^*| < \delta$ implies $|f(x) - f(c^*)| < 1$. By the triangle inequality, also

$$||f(x)| - |f(c^*)|| < 1,$$

and therefore,

(A) $$|f(x)| < |f(c^*)| + 1.$$

Because $c^* - \delta < c^*$, $c^* - \delta$ is not an upper bound of C and there is a number $t \in C$ such that $c^* - \delta < t \leqslant c^*$. See Figure 5.1. Also, if w is in $[a,b]$ with $c^* \leqslant w < c^* + \delta$, then $x \in [t,w]$ implies $|x - c^*| < \delta$, so that for these numbers x, line (A) is true also,

$$|f(x)| < |f(c^*)| + 1.$$

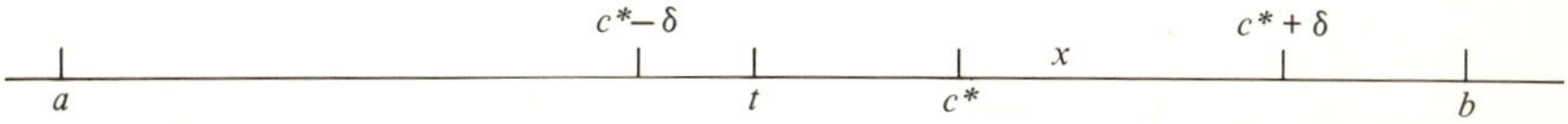

Figure 5.1

By the definition of C there is a number M_t that bounds $|f(x)|$ on $[a,t]$, and we have just seen that $|f(c^*)| + 1$ bounds $|f(x)|$ on $[t,w]$; therefore,

$$|f(x)| \leqslant M_t + |f(c^*)| + 1, \quad x \in [a,w].$$

Two things are now clear. First, $c^* \in C$ because $c^* \in [a,w]$, and second, $c^* \neq \text{ub } C$ if $w > c^*$. Therefore, there is no number in $[a,b]$ larger than c^*, and $c^* = b$. As we observed above, this says that f is a bounded function.

So range f is a nonempty, bounded set of numbers, and the least upper bound axiom applied to range f provides a number $N = \text{lub range} f$. When we establish that $N \in \text{range } f$; that is, that $f(c) = N$ for some $c \in \text{dom} f$, then we will have

$$f(x) \leqslant N = f(c), \quad x \in \text{dom} f$$

and $f(c)$ will be the absolute maximum of f.

Suppose that there were no such number c, and therefore that $N - f(x) > 0$, $x \in \text{dom} f$. Then $1/(N - f(x))$ describes a continuous function whose domain is a closed interval; namely, dom f, and, by the argument we just made, it is bounded by a positive number, say $M > 0$.

$$\frac{1}{N - f(x)} \leqslant M, \quad x \in \text{dom} f$$

Therefore,

$$f(x) \leqslant N - \frac{1}{M}, \quad x \in \text{dom} f,$$

that is,

$$N > N - \frac{1}{M} = \text{ub range} f,$$

contradicting $N = \text{lub range } f$. Therefore there is some $c \in \text{dom } f$ for which

$$f(x) \leqslant f(c), \quad x \in \text{dom } f.$$

Now modify this argument to show that f has an absolute minimum also. ■

5.2 PROBLEMS

1. Show that

$$f(x) = \frac{1}{x}, \quad x > 0$$

has no absolute extreme value. Explain why this does not conflict with the extreme value lemma.

2. Proceed as in Problem 1 for each of the following.
   (a) $f(x) = 1/x, \quad -2 \leqslant x \leqslant 2, \quad x \neq 0$
   $ = 4 \quad , \qquad\qquad\qquad x = 0$
   (b) $f(1/n) = (-1)^n n, \quad n = 1,2,3, \ldots$
   $f(x) = 0 \qquad\quad , \quad$ otherwise on $[0,1)$
   (c) $f(x) = x^2, \quad 0 < x < 1$
3. Prove that if $\text{dom } f = [a,b]$ and $f(x) = P(x)$, $x \in \text{dom } f$, where P is a polynomial function, then f has an absolute maximum and minimum.
4. Prove that if

$$\begin{aligned} f(x) &= |x|, \quad -1 \leqslant x \leqslant 1, \quad x \neq 0 \\ &= -1, \qquad\qquad\qquad\quad x = 0, \end{aligned}$$

then $\text{dom } f$ is a closed interval, f is not continuous, but f has both absolute extremes. Does this conflict with the extreme value lemma?

5. Consider

$$f(x) = \sin x, \quad 0 < x < 2\pi.$$

Here, $\text{dom } f$ is not a closed interval, but f takes both absolute extremes. Does this conflict with the extreme value lemma?

6. Construct an example of a discontinuous function on a nonclosed interval which has both absolute extremes.

Knowing that a continuous function on a closed interval always has extreme values, we can prove a very useful property of differentiable functions on closed intervals. This is formulated as Theorem 5.1 below.

Lemma 1 to Theorem 5.1: Let $f(c)$ be a local extreme of a differentiable function f, where c is interior[2] to $\text{dom } f$. Then $Df(c) = 0$.

[2] There is an open interval I, containing c, so that $I \subset \text{dom } f$.

Proof: Recall that $Df(c) = \lim_{0} {}_f\sigma_c$. Therefore, if $Df(c) > 0$, then by Theorem 3.10 there is a number $\delta > 0$ so that

$$\frac{1}{h}\,[f(c+h) - f(c)] > 0, \quad 0 < |h| < \delta.$$

For $h > 0, f(c+h) > f(c)$, and for $h < 0, f(c+h) < f(c)$ so $f(c)$ cannot be a local extreme. Similarly, if $Df(c) < 0, f(c)$ cannot be a local extreme. Therefore, $Df(c) = 0$. ■

We have noticed that the graph of a differentiable function appears to be a smooth curve without corners. Such a graph could appear as in Figure 5.2. At some number between a and b the tangent line should be horizontal. The next lemma verifies this.

Figure 5.2

ROLLE'S LEMMA If f is a differentiable function whose domain is a closed interval $[a,b]$, and if $f(a) = f(b) = 0$, then there is a number $c \in (a,b)$ at which $Df(c) = 0$.

Proof: Suppose that f is a constant function

$$f(x) = 0, \quad x \in [a,b].$$

Then $Df(x) = 0$ at each $x \in [a,b]$.

Otherwise, suppose that $f(x) > 0$ for some $x \in (a,b)$. Now f is a continuous function, necessarily, and by the extreme value lemma has an absolute maximum, say $f(\bar{c}) > 0$. Thus, $\bar{c} \neq a$, $\bar{c} \neq b$. So $\bar{c} \in (a,b)$ and by Lemma 1, $Df(\bar{c}) = 0$.

If f is neither the constant 0 function nor takes a positive value, then $f(x) < 0$ for some $x \in (a,b)$, and has an absolute minimum, say $f(\underline{c}) < 0$, and again $\underline{c} \in (a,b)$ and $Df(\underline{c}) = 0$. ■

Rolle's lemma is generalized easily to the following theorem.

THEOREM 5.1 (mean value theorem) Let f be a differentiable function whose domain is a closed interval $[a,b]$. Then there is a number $c \in (a,b)$ at which

$$f(b) = f(a) + (b - a)\, Df(c).$$

Proof: We get a hint from Figure 5.3. P is a point on the graph of f, and Q is the intersection of the vertical line through P with the chord

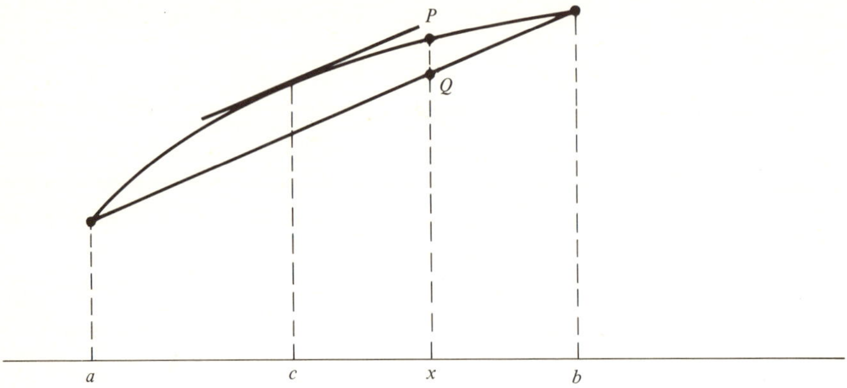

Figure 5.3

joining the end points of the graph of f. Construct the function g whose domain is dom f and whose value measures the length of the segment $\overline{PQ}$.

$$g\colon g(x) = f(x) - f(a) - \frac{f(b) - f(a)}{b - a}(x - a), \quad x \in \operatorname{dom} f.$$

Notice that g is differentiable, dom g is a closed interval, and $g(a) = g(b) = 0$. In particular,

$$Dg(x) = Df(x) - \frac{f(b) - f(a)}{b - a}, \quad x \in \operatorname{dom} f.$$

Rolle's lemma provides a number $c \in (a,b)$ at which $Dg(c) = 0$, and this gives the formula of the theorem. ■

If the formula of Theorem 5.1 is written as

$$Df(c) = \frac{f(b) - f(a)}{b - a}$$

and $Df(c)$ is interpreted as the slope of the tangent line to the graph of f at $P(c,f(c))$, then this tangent line is parallel to the chord of Figure 5.3. If you recall the proof of Rolle's lemma you will see that the number c determines where $\overline{PQ}$ is longest. Of course, the formula of the theorem may also be true for other numbers than this particular c.

With the mean value theorem we can prove a restricted converse of Theorem 4.1.

THEOREM 5.2 Let g be a differentiable function whose domain is the closed interval $[a,b]$. If $Dg = 0$, then g is a constant function.

Proof: For any number $x \in [a,b]$, g is differentiable on $[a,x]$,[3] and

[3] In other words, the restriction of g to the interval $[a,x]$ is a differentiable function.

the mean value theorem provides a number $c \in (a,x)$ at which

$$g(x) = g(a) + (x - a)\, Dg(c) = g(a). \blacksquare$$

Notice that Theorem 5.2 does not pick out which constant function g is; it only asserts that whatever value g takes at a, it keeps throughout its domain.

Given a function and its derivative function, we ask what other function(s) could have this same derivative.

Corollary to Theorem 5.2: Let the functions g and h be differentiable with the common domain $[a,b]$. If $Dg = Dh$, there is a constant function k so that $g = h + k$.

Proof: Observe that $D(g - h) = Dg - Dh = 0$. Therefore, by Theorem 5.2

$$\begin{aligned}(g-h)(x) &= (g-h)(a)\\ g(x) &= h(x) + (g-h)(a),\end{aligned}$$

and we take $k(x) = (g - h)(a), \quad a \leqslant x \leqslant b.$ $\blacksquare$

5.2 PROBLEMS (*continued*)

7. Illustrate Lemma 1 to Theorem 5.1 with the function

$$f(x) = x(x-3)^2, \quad 0 \leqslant x \leqslant 4$$

as follows.
   (a) Compute $f(0 + h) = h(h - 3)^2$, $h > 0$, and deduce that 0 is a local minimum at 0. Compute $Df(0)$.
   (b) Compute $f(1 + h) = 4 - h^2\,(3 - h)$, and deduce that 4 is a local maximum at 1. Compute $Df(1)$.
   (c) Compute $f(3 + h) = 0 + h^2\,(3 + h)$, and deduce that 0 is a local minimum at 3. Compute $Df(3)$.
   (d) Compute $f(4 + h) = 4 + h\,(3 + h)^2$, $h < 0$, and deduce that 4 is a local maximum at 4. Compute $Df(4)$.
8. Let $f(x) = x(x - 3)^2$, $0 \leqslant x \leqslant 3$. Find the number c of Rolle's lemma.
9. Let $f(x) = x(x - 3)^2$, $0 \leqslant x \leqslant 4$. Find a number c of the mean value theorem. Find another.
10. Prove the following theorem.

THEOREM Let f be a continuous function with dom $f = [a,b]$, and $f(a) = f(b) = 0$. If f is differentiable on (a,b), then there is a number $c \in (a,b)$ at which $Df(c) = 0$.

11. You may assume that sin, on any interval, is differentiable, and that

$D\sin = \cos$. Consider

$$f(x) = x \sin \frac{1}{x}, \quad 0 < x \leqslant \frac{1}{2\pi}$$
$$= 0 \qquad , \quad x = 0.$$

Verify that f does not satisfy the hypothesis of Rolle's lemma, but that it does satisfy the hypothesis of the theorem of Problem 10. Find several numbers c at which $Df(c) = 0$.

12. Using the theorem of Problem 10, prove the following theorem.

THEOREM Let f be a continuous function with dom $f = [a,b]$, and be differentiable on (a,b). Then there is a number $c \in (a,b)$ at which $f(b) = f(a) + (b - a)\,Df(c)$.

13. Let $f(x) = 1 - |x - 1|$, $0 \leqslant x \leqslant 2$. Verify that $f(0) = f(2) = 0$ but that there is no number $c \in (0,2)$ at which $Df(c) = 0$. Does this contradict Rolle's lemma?
14. Show that dom $f = [0,2]$ where

$$f(x) = 2 - (x-1)^2, \quad 0 < x < 1$$
$$= 0 \qquad , \quad x = 0, \quad x = 2$$

and that f is neither continuous nor differentiable. Compute $Df(1)$. Does this conflict with Rolle's lemma?

15. Let f be a twice differentiable function whose second derivative D^2f is absolutely bounded by M. Suppose that we know $f(x)$, and we compute $f(x + h)$, $h \neq 0$ by the formula

$$f(x) + df(x,h).$$

Then the error is

$$_xe(h) = f(x+h) - f(x) - df(x,h).$$

Prove that $|_xe(h)| < Mh^2$.
Hint: By the mean value theorem for f, there is a number θ such that

$$f(x+h) - f(x) = Df(x + \theta h)h, \quad 0 < \theta < 1,$$

so that

$$e(h) = [Df(x + \theta h) - Df(x)]h.$$

Apply the mean value theorem again to Df, and exhibit a number ϕ such that

$$Df(x + \theta h) - Df(x) = D^2f(x + \phi\theta h)\theta h, \quad 0 < \phi < 1.$$

5.3 EXTREME VALUES OF A DIFFERENTIABLE FUNCTION

We are now in a position to initiate a search for the local extremes of a differentiable function. By Lemma 1 to Theorem 5.1 we know that $Df(c) = 0$ at any local extreme c interior to dom f. Therefore such local extremes will be found among the numbers $\{f(c) | Df(c) = 0\}$. In order to sort these out we appeal to the next theorem.

THEOREM 5.3 Let f be a differentiable function, and let c be a number interior to dom f. In each of the cases (a), (b), and (c), suppose there is an open interval $I \subset$ dom f, which contains c, and let $x \in I$.

(a) If $Df(x) > 0$ for all $x < c$, and $Df(x) < 0$ for all $x > c$, then $f(c)$ is a local maximum at c.
(b) If $Df(x) < 0$ for all $x < c$, and $Df(x) > 0$ for all $x > c$, then $f(c)$ is a local minimum at c.
(c) If either $Df(x) > 0$ for all $x \neq c$, or $Df(x) < 0$ for all $x \neq c$, then $f(c)$ is not a local extreme at c.[4]

Proof: Our statement of the mean value theorem referred to a differentiable function per se, and in this proof we will apply the theorem to f over only a portion of its domain. This is why it will be necessary to introduce auxiliary functions ${}_xf_c$ and ${}_xg_c$, which are restrictions of f to the intervals $[x,c]$ and $[c,x]$. With the understanding that this restriction of f can be done, the proof can be condensed. First we will prove case (a), and then case (c) in condensed form.

Case (a): For any $x \in I$, with $x < c$ define the function ${}_xf_c$ by ${}_xf_c(t) = f(t)$, $x \leqslant t \leqslant c$ (see Figure 5.4); then ${}_xf_c$ is a differentiable function and by the mean value theorem there is a number $c' \in (x,c)$ at which

$$\text{(A)} \qquad \begin{gathered} {}_xf_c(c) = {}_xf_c(x) + (c-x)\, D_x f_c(c'), \quad \text{and} \\ (c-x)\, D_x f_c(c') > 0, \quad \text{because } D_x f_c(c') = Df(c') > 0. \end{gathered}$$

For any $x \in I$, with $x > c$, define the function ${}_xg_c$ by

$${}_xg_c(t) = f(t), \quad c \leqslant t \leqslant x.$$

Then, as above, there is a $c'' \in (c,x)$ at which

$$\text{(B)} \qquad \begin{gathered} {}_xg_c(x) = {}_xg_c(c) + (x-c)\, D_x g_c(c''), \quad \text{and} \\ (x-c)\, D_x g_c(c'') < 0, \quad \text{because } D_x g_c(c'') = Df(c'') < 0. \end{gathered}$$

[4] In case (c), if $Df(c) = 0$, the point $P(c, f(c))$ is called a *horizontal point of inflection* of the graph of f.

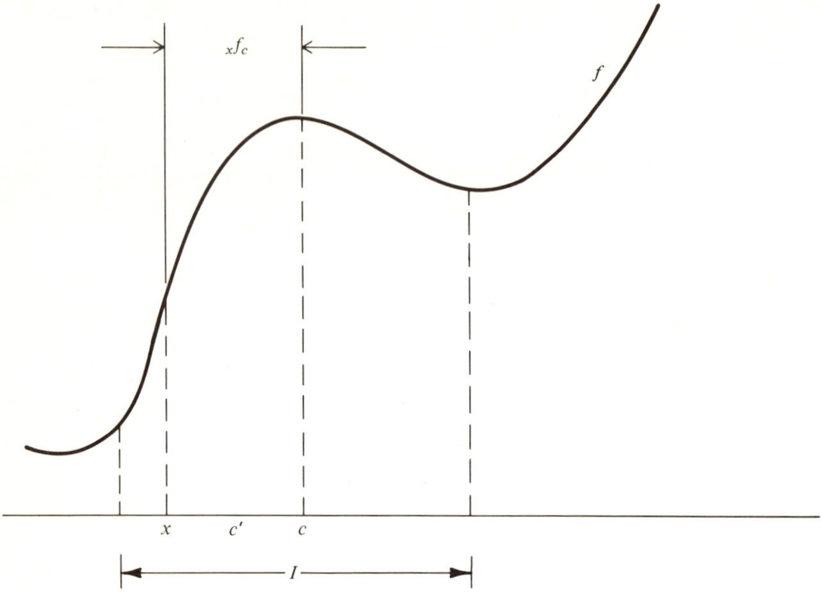

Figure 5.4

Therefore, for $x \in I$, if $x < c$, line (A) implies

$$f(x) = {}_xf_c(x) < {}_xf_c(c) = f(c),$$

and for $x > c$, line (B) implies

$$f(x) = {}_xg_c(x) < {}_xg_c(c) = f(c),$$

and so $f(c)$ is a local maximum at c.

Case (c) ($f(x) > 0$): On the interval $[x,c]$, f is a differentiable function, and there is a number $c' \in (x,c)$ at which

$$f(c) = f(x) + (c - x)\ Df(c'),$$

and $(c - x)\ Df(c') > 0$. See Figure 5.5. Also on the interval $[c,x]$ there is a number $c'' \in (c,x)$ at which

$$f(x) = f(c) + (x - c)\ Df(c'')$$

and $(x - c)\ Df(c'') > 0$.

Therefore, for $x \in I$, if $x < c$

$$f(x) < f(c),$$

and if $x > c$,

$$f(x) > f(c),$$

and so $f(c)$ is not a local extreme at c. ■

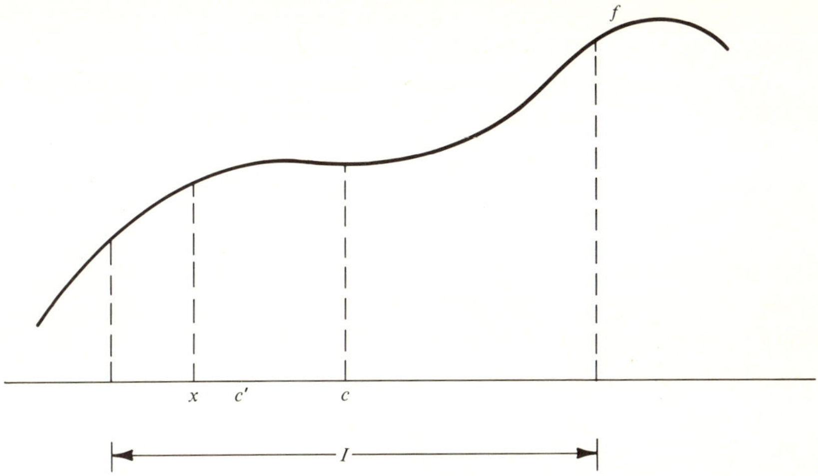

Figure 5.5

In view of Lemma 1 to Theorem 5.1, the extremes of f referred to in Theorem 5.3 can occur only at numbers c for which $Df(c) = 0$. Hence, we find these extremes by solving $Df(c) = 0$ for c, and then Theorem 5.3 will help to sort out the types of extremes.

EXAMPLE See Figure 5.6.

$$f(x) = x^3(x+2)^2, \quad x \text{ real}$$
$$Df(x) = x^2(x+2)(5x+6)$$
$$Df(c) = 0 \text{ for } c = 0, -2, -\tfrac{6}{5}$$

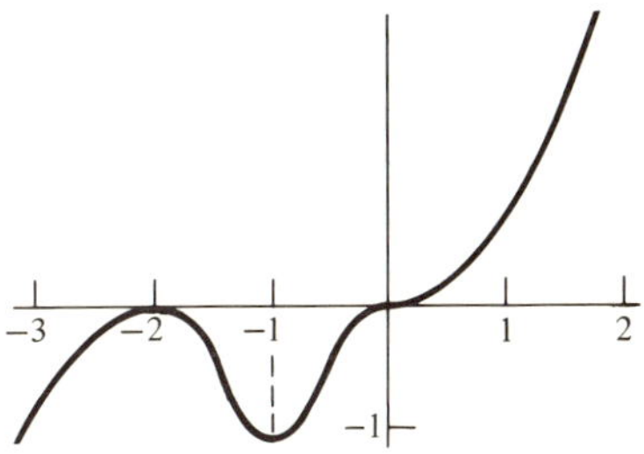

Figure 5.6

Apply Theorem 5.3 as follows: At $c = -2$ take $I = (-\frac{5}{2}, -\frac{3}{2})$; then $Df(c) > 0$, $x \in (-\frac{5}{2}, -2)$ and $Df(x) < 0$, $x \in (-2, -\frac{3}{2})$, so $f(c) = f(-2) = 0$ is a local maximum at -2.

At $c = -\frac{6}{5}$ take $I = (-\frac{7}{5}, -1)$; then $Df(x) < 0$, $x \in (-\frac{7}{5}, -\frac{6}{5})$ and $Df(x) > 0$, $x \in (-\frac{6}{5}, -1)$, so $f(c) = f(-\frac{6}{5}) = -3456/3125$ is a local minimum at $-\frac{6}{5}$.

At $c = 0$ take $I = (-1,1)$; then $Df(x) > 0$, $x \in (-1,0)$ and $Df(x) > 0$, $x \in (0,1)$; so $f(c) = f(0) = 0$ is not a local extreme at 0.

In this example every number is interior to dom f, and therefore, by Lemma 1 to Theorem 5.1 there are no other local extremes.

It happens in this example that Df is itself a differentiable function, a polynomial function, and in such cases the identification of the kinds of extremes can sometimes be shortened.

THEOREM 5.4 Let f be a differentiable function whose derivative, Df, is also differentiable. Suppose that $f(c)$ is an extreme value of f at c, where c is interior to dom f. Then if $D^2f(c) > 0$, $f(c)$ is a local minimum at c, and if $D^2f(c) < 0$, $f(c)$ is a local maximum at c.

Proof: Since c is interior to dom Df and, for the case $D^2f(c) > 0$,

$$D^2f(c) = \lim_{h \to 0} \frac{1}{h}\,[Df(c+h) - Df(c)] > 0,$$

there is a number $\delta > 0$ such that

$$\frac{1}{h}\,[Df(c+h) - Df(c)] = \frac{1}{h}\,Df(c+h) > 0, \quad 0 < |h| < \delta.$$

Let $x = c + h$; then if $x < c$, $h < 0$ and

$$Df(x) = Df(c+h) < 0,$$

while if $x > c$, $h > 0$, then

$$Df(x) = Df(c+h) > 0.$$

By Theorem 5.3, $f(c)$ is a local minimum at c. The case $D^2f(c) < 0$ is easily handled similarly. ■

EXAMPLE (continued)

$f(x) = x^3(x+2)^2$

$Df(c) = 0$ at $c = 0, -2, -\frac{6}{5}$

$D^2f(x) = 4x(5x^2 + 12x + 6)$

$D^2f(-2) = -16 < 0$, $f(-2)$ is a local maximum at -2

$D^2f(-\frac{6}{5}) = 144/25 > 0$, $f(-\frac{6}{5})$ is a local minimum at $-\frac{6}{5}$

$D^2f(0) = 0$, a case not covered by Theorem 5.4.

5.3 PROBLEMS

Each of the functions in Problems 1, 2, 3, and 4 is differentiable. Find the extreme values at numbers interior to the domains.

1. (a) $f(x) = x^3 - x^2 - 8x + 6$

(b) $f(x) = x^2 - 4x + 1$
(c) $f(x) = (x-1)^2(x+2)$
(d) $f(x) = (x-1)^3$
(e) $f(x) = (x-1)^3(x+1)^2$
(f) $f(x) = 1 - 7x + 3x^2$
(g) $f(x) = 1 - 7x - 3x^2$
(h) $f(t) = 2t^3 + 3t^2 - 36t$
(i) $g(y) = 2y^3 - 3y^2 + 6y$
(j) $\rho(t) = t^3 - 6t^2 + 9t + 12$
(k) $g(y) = y^4 - 4$
(l) $g(s) = s^4 - 4s^3 + 4s^2 - 4$
(m) $\phi(x) = x^4 - 2x^2 + 4$
(n) $g(t) = t^4 - 4t + 4$
(o) $f(z) = az^2 + bz + c$

2. (a) $f(x) = x^3 + 3/x, \quad x > 0$
(b) $f(x) = x^3 + 3/x, \quad x < 0$
(c) $f(x) = x^2 + 1/x, \quad x > 0$
(d) $f(x) = x^2 + 1/x, \quad x < 0$
3. (a) $f(x) = (x^2+1)/x, \quad x > 0$
(b) $f(x) - (x^2+1)/x, \quad x < 0$
(c) $f(x) = x/x^2 + 1$
(d) $f(x) = x^2/x + 1, \quad x < -1$
(e) $f(x) = x^2/x + 1, \quad x > -1$
4. (a) $f(x) = x(x+1)^{1/3}, \quad x > -1$
(b) $f(x) = x^{2/3}(x+1)^{1/3}, \quad -1 < x < 0$
(c) $f(x) = x^{1/3}(x+1)^{2/3}, \quad -1 < x < 0$
(d) $f(x) = x^{2/3}(x+1)^{2/3}, \quad -1 < x < 0$
5. Prove the following theorem.

THEOREM Let f be a differentiable function with dom $f = [a,b]$. If $Df(a) > 0$, then $f(a)$ is a local minimum at a.

6. State and prove three more theorems like the theorem of Problem 5.
7. Prove the following theorem.

THEOREM Let f be a continuous function with dom $f = [a,b]$, and suppose that f is differentiable on (a,b) with $Df(a) = 0$ if it exists. If there is a number $c \in (a,b)$ such that $Df(x) > 0$ for $a < x < c$, then $f(a)$ is a local minimum at a.

8. Prove three more theorems like the theorem of Problem 7.
9. Find the extreme values of the following.
(a) $f(x) = \sqrt{x}, \quad x \geqslant 0$
(b) $f(x) = x^{3/2}, \quad x \geqslant 0$
(c) $f(x) = \sqrt{1+x}, \quad x \geqslant -1$

(d) $f(x) = \sqrt{1-x}, \quad x \leqslant 1$
(e) $f(x) = \sqrt{1-x^2}, \quad |x| \leqslant 1$
(f) $f(x) = (1-x^2)^{3/2}, \quad |x| \leqslant 1$
(g) $f(x) = (x^2-1)^{3/2}, \quad x \geqslant 1$
(h) $f(x) = (x^2-1)^{3/2}, \quad x \leqslant -1$

5.4 THE GENERAL EXTREME VALUE PROBLEM

Let us devote our attention to a function f whose domain is an interval. Then we ask whether the function takes an extreme value at a number $c \in \operatorname{dom} f$. The following cases are exhaustive alternatives.

(a) c is not interior to dom f.
(b) c is interior to dom f and $Df(c)$ does not exist.
(c) c is interior to dom f and $Df(c)$ exists.

In Section 5.3 we developed some techniques for discovering and classifying extremes under case (c). For cases (a) and (b) we are left with a direct examination of f, and we must rely on the initial definitions of extreme values or on Theorem 5.3 and those theorems of Problems 5, 6, 7, and 8 when they apply.

EXAMPLE $f(x) = 4(x-1)^2, \quad 0 \leqslant x \leqslant 3$
$ = 64 - 16x, \quad 3 < x < 4$

The number $c = 0$ is not interior to dom f (case (a)). On the interval I, $|x| < 1$, $x \in I \cap \operatorname{dom} f$ implies that

$$f(x) = 4(x-1)^2 < 4 = f(0)$$

and that $4 = f(0)$ is a local maximum at 0.

If $0 < c < 3$, c is interior to dom f and $Df(c)$ exists (case (c)). Therefore the analysis of Section 5.3 applies, and $Df(c) = 0$ for $8(c-1) = 0$; that is, $c = 1$. Since $D^2f(c) = 8 > 0$, $0 = f(1)$ is a local minimum at 1.

If $c = 3$, c is interior to dom f and $Df(c)$ does not exist (case (b)). On the interval I, $|x-3| < 1$, $x \in I \cap \operatorname{dom} f$ implies that

$$f(x) = 4(x-1)^2 < 16 = f(3), \quad x < 3$$
$$f(x) = 64 - 16x < 16 = f(3), \quad x > 3$$

and that $16 = f(3)$ is a local maximum at 3.

If $3 < c < 4$, c is interior to dom f and $Df(c)$ exists (case (c)). Therefore, since $Df(c) = -16 \neq 0$, there is no local extreme.

If $c \in \operatorname{dom} f$, then c falls within one of these alternatives and we have discovered all of the extreme values of f. Notice that although

$f(2) = 4$ is a local extreme, it is not a local extreme at 2. Both absolute extremes exist: 16 is the absolute maximum and 0 is the absolute minimum.

5.4 PROBLEMS

1. Examine for extreme values.
   (a) $f(x) = \dfrac{x^2 + 2x - 23}{x - 4}$
   (b) $h(x) = \dfrac{1 + x + x^2}{1 - x + x^2}$
   (c) $\phi(x) = \dfrac{1 - x + x^2}{1 + x + x^2}$
   (d) $h(y) = \dfrac{y^2 - 1}{1 + y + y^2}$
   (e) $g(t) = \dfrac{1}{t} + \dfrac{1}{1 - t}$
2. (a) $f(x) = x^{1/3}$
   (b) $f(x) = x^{2/3}$
   (c) $f(x) = (x - 1)^{1/5}$
   (d) $f(x) = |x|$
3. (a) $f(x) = x^4 - 6x^2 + 8x + 2$
   (b) $f(x) = x^2 - (2/x)$
   (c) $f(x) = x\sqrt{1 - x^2}$
   (d) $f(x) = x\sqrt{x - x^2}$
4. The sum of two numbers is 10. Find these numbers if their product is as large as possible.
5. The sum of two numbers is 10. Find these numbers if the sum of their squares is as small as possible.
6. Find the positive number such that the sum of the number and its reciprocal is as small as possible.
7. Show that the sum of the square of any number and the square of its reciprocal is always greater than or equal to 2.
8. Of all rectangles with a given perimeter, find the one with the largest area.
9. The equal sides of an isosceles triangle are 10 inches long. What is the length of the base of the largest such triangle?
10. In Problem 5, Section 2.5, find the dimensions of the largest box.
    Answer: $x = 5 - (5\sqrt{3})/3$ inches
11. In Problem 7, Section 2.5, find the dimensions of the cheapest box.
12. In Problem 9, Section 2.5, find the area of the largest window.
13. In Problem 13, Section 2.5, is there a largest or smallest triangle?
14. An isosceles triangle has a base 10 units long and an altitude 20 units

high. Rectangles are inscribed, with one side on the base of the triangle. What are the dimensions of the largest such rectangle?

15. Rectangles are inscribed in a circle of radius R. Find the dimensions of the rectangle whose perimeter has an extreme value. Is this extreme a relative maximum or minimum?
    Answer: The square with side $R\sqrt{2}$ has a maximum area.
16. Find the dimensions of the largest right circular cylinder which can be cut from a solid sphere of radius R.
17. Find the dimensions of the largest right circular cylinder which can be cut from a solid right circular cone of height H and radius of base R.
18. A cylindrical tank without a top is to have a volume of A cubic feet. Show that for a minimum amount of metal, the radius of the base should equal the altitude.
19. Find the shortest segment which has its ends on the axes and is tangent to the graph of the equation $xy = 1$.
20. A tank has the form of a cylinder with hemispherical ends. If the volume is A cubic feet, what are its dimensions for a minimum amount of material?
    Answer: Radius of end is $(3A/4\pi)^{1/3}$.
21. A piece of wire 1 foot long is cut into two pieces. One is bent into a square and the other into a circle. Where should the cut be made if the sum of the areas of the square and circle is to be an extreme? Which of these extremes are relative maxima and which are relative minima?
22. What point on the curve $46 = x^2$ is nearest to the point (0,4)?
23. On the circle whose equation is $x^2 + y^2 = 25$ find the point nearest to the point (6,8).
24. A rectangular poster is to contain 24 square inches of printed matter and is to have a margin at top and bottom of 2 inches and a margin at the sides of 1 inch. What are the dimensions of the poster using the least amount of paper?
    Answer: $2\sqrt{3}$ by $4\sqrt{3}$ inches
25. The demand for a certain article varies inversely as the cube of the selling price. If the articles cost 20 cents apiece to manufacture, find the selling price which yields a maximum profit.
26. A taxi company charges 15 cents a mile and logs 600 passenger miles a day. Twenty-five fewer passenger miles a day would be logged for each cent increase in the rate per mile. What rate yields the greatest gross income?
27. In Problem 26 if the operating costs are 10 cents per mile, what rate yields the greatest net income?
28. A man is in a boat 1 mile from the nearest point A of a straight shore. He wishes to arrive as soon as possible at a point B 3 miles along the

shore from A. He can row 2 miles an hour and walk 4 miles an hour. Where should he land?
Answer: $\sqrt{3}/3$ miles from A

29. How should he proceed in the above problem if he can row 5 miles an hour and walk 4 miles an hour?
30. A rectangular corral is to be built, using one side of a barn B feet long, and L feet of fencing. What are the dimensions of the largest possible corral?
31. Along a street 3 miles long potential customers are evenly distributed. One mile from one end of the street is a dime store. A new dime store is to be located on this street. If each customer will trade at the store nearest him, is there a best location for the new store?

REVIEW PROBLEMS

1. A function f, defined only on the interval $[a,b]$, has for its graph Figure 5.7.

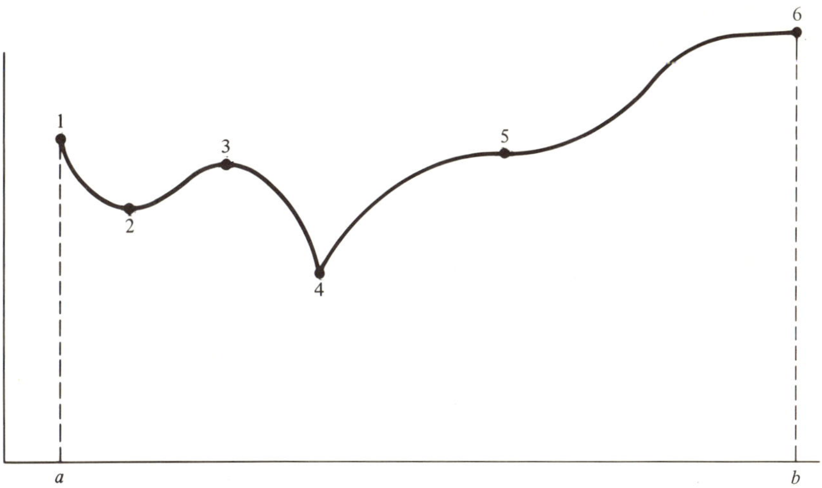

Figure 5.7

(a) Which of the six points is
  (i) an absolute maximum?
  (ii) an absolute minimum?
  (iii) a relative maximum?
  (iv) a relative minimum?

(b) What can you say about the value of Df and D^2f at each of these six points?

2. For which values of a and b will the graph of the function h have a

horizontal tangent at (1,5) if

$$h(t) = at^2 + \frac{b}{t^3}?$$

Does h have a relative maximum or minimum at (1,5)? Are there any other extremes of h?

3. A function g is such that D^2g is continuous on the interval $[a,b]$. The equation $g(x) = 0$ has three different solutions in the open interval (a,b). Show that the equation $D^2g(x) = 0$ has at least one solution in (a,b).
   Hint: Use Rolle's lemma.
4. Show that $x^2 + x + 1$ is positive for all values of x.
5. Find the greatest and least values of the fraction whose denominator is 49 greater than the square of its numerator.
   Answer: $-7/98$ least, $7/98$ greatest
6. What number exceeds its kth power by the greatest amount, k being a positive integer?
7. Find the maximum area of an isosceles triangle whose perimeter is 6 inches.
8. In a triangle ABC the angle A is fixed while the side BC passes through a given point D lying within the angle A. Show that the area of the triangle is a minimum when D is the midpoint of BC.
9. Through the point (a,b) a line is drawn such that the part intercepted between the axes is a minimum. Prove that its length is

$$(a^{2/3} + b^{2/3})^{3/2}.$$

10. If the radius of a sphere is 12 inches and is increasing at the rate of $\frac{1}{12}$ inches per second, at what rate is the volume of the maximum inscribed right circular cylinder increasing?
11. Water flows out of a hemispherical basin through a hole at the bottom so that the volume of the water remaining at any time decreases at a rate proportional to the square root of the depth of the water remaining. Prove that the level of the water falls most slowly when the depth is two-thirds of the radius of the basin. The volume of a spherical segment of one base is

$$\frac{\pi}{3}(3Rh^2 - h^3),$$

where R is the radius of the sphere and h is the height of the segment.

12. The segment of the graph of $y^2 = 2px$ bounded by the line $x = p/2$ is rotated about the x-axis forming a solid of revolution. Find the dimensions of the cylinder of maximum volume which can be inscribed in the solid.
13. A wall 27 feet high is 8 feet from a building. Find the length of the

shortest ladder which will reach the building if one end rests on the ground outside of the wall.

14. A girder 27 feet long is moved on rollers along a passageway and into a corridor 8 feet wide at right angles to the passageway. Neglecting the horizontal width of the girder, find how wide the passageway must be in order that the girder may go around the corner.
Answer: $5\sqrt{5}$ feet
15. Cylindrical cans, with top and bottom, are to be made. The top and bottom pieces are cut from square pieces of metal and the scraps left over from these squares are wasted. There is no other waste. Find the ratio of height-to-diameter of the least expensive can.
16. Find the area of the largest rectangle with the lower base on the x-axis and the upper vertices on the curve $y = 1 - x^2$.
17. A line segment has one end on the line whose equation is $y = -1$. It passes through the origin and is terminated at the point where it meets the curve $y = x^2$ a second time. Find the slope of the shortest segment which can be drawn.
Answer: $\pm 1\sqrt{2}$
18. Find the point on the graph of $y = \sqrt{x}$ which is nearest to
(a) (1,0)
(b) $(\frac{1}{4},0)$
(c) $(-1,0)$.
19. At a point P in the first quadrant on the curve $y = 7 - x^2$ a tangent is drawn, meeting the coordinate axes at A and B. Find the position of P which makes AB a minimum.
20. Show that for a fixed a, the function

$$y = \left(a - \frac{1}{a} - x\right) \cdot (4 - 3x^2)$$

has just one maximum and one minimum, and that the difference between them is

$$\frac{4}{9}\left(a + \frac{1}{a}\right)^3.$$

If a is allowed to vary, find the least value of this difference.

21. If f is a function which has a second derivative at each point of an interval $[a,b]$, show that there is a number d inside this interval such that

$$f(b) = f(a) + Df(a)(b - a) + \frac{D^2f(d)}{2}(b - a)^2.$$

Hint: Consider the function ϕ defined by

$$\phi(x) = f(x) - f(a) - Df(a)(x - a) - k(x - a)^2,$$

where the number k is so chosen that $\phi(b) = 0$.

22. DOUBLE MEAN VALUE THEOREM
If f and g are functions which are differentiable in an interval $[a,b]$, and if $Dg(x) \neq 0$ for any x in $[a,b]$, show that there is a number c, with $a < c < b$, such that

$$\frac{f(b) - f(a)}{g(b) - g(a)} = \frac{Df(c)}{Dg(c)}.$$

Hint: Consider the function ϕ defined by

$$\phi(x) = \frac{f(b) - f(a)}{g(b) - g(a)} (g(x) - g(a)) - (f(x) - f(a)).$$

Chapter 6 THE INTEGRAL OF A FUNCTION

6.1 INTRODUCTION

Let us reflect briefly on a few of the problems that have given direction to our discussions thus far. Earlier, it appeared to be very desirable that the number system be complete; that is, that sequences of approximations of increasing accuracy lead to answers. Without the property of completeness, introduced by the least upper bound axiom, we could, for example, formulate a process for the solution of $x^2 = 2$ which has arbitrarily fine accuracy, but which converges on nothing at all. Think back on what has been done, and you will recall that often the existence of a number needed, such as the extreme value of a function, rested squarely on this completeness of the real number system.

As we began to explore properties of functions we found such a bewildering population of them that it was necessary to focus on essentially just two of these properties. For practical, and indeed aesthetic, reasons unbrokenness was singled out first of all; we called it continuity.

This is related to what we feel to be the uninterrupted flow of time, or continuous motion. Our resources for dealing with this motion are limited essentially to finite, discrete steps, say, the number of lines in an "$\epsilon - \delta$" proof. The tool used to examine continuity was the limit of a function. The other property we investigated was smoothness, or lack of corners. We discovered that the slope function, combined with the limit to create a derivative, could sense this smoothness or lack of it.

The use of the complete real number system to model the "time axis" of physics has led to extensive use of the calculus for the analysis and description of physical phenomena which are related to the flow of time. Thus smooth motion has become identified with a differentiable function and vice versa.

Also we commented on a way to find the area of a circle; more generally, the area of a figure not made up of triangles. This simple geometrical problem has surprising generality, but an attack on it has an awkward beginning. The slope function was an entering wedge into a study of smoothness by means of the derivative. Now, a vast collection of sums of terms will be an entering wedge into a study of areas and other things like them, leading to the definite integral. We will shortly show that these two very disparate wedges point toward a close relationship between the derivative and the integral, and we will state this as the fundamental theorem in this chapter.

Suppose that we examine two problems which introduce these sums: one geometrical for the geometers, the other physical for the physicists, and both, we trust, of interest to you.

Look at the quarter circle in Figure 6.1. It is the graph of the function

$$f: f(x) = \sqrt{1 - x^2}, \quad 0 \leqslant x \leqslant 1.$$

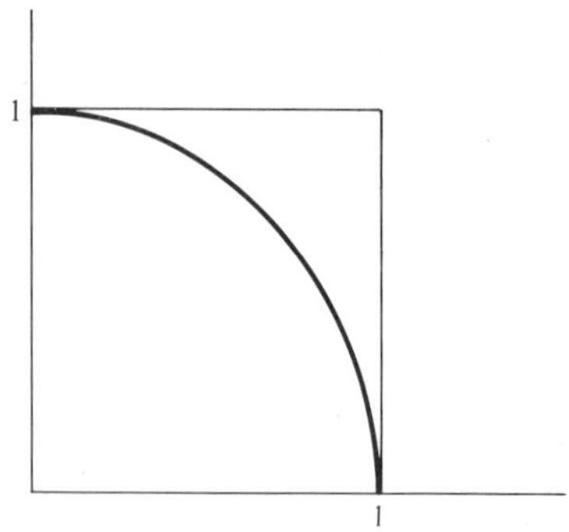

Figure 6.1

Its area A ought to be less than the area of the unit square and greater than 0

$$0 \leq A \leq 1,$$

pretty rough. However, if the figure is cut into vertical slices, as in Figure 6.2, each slice of the circle being viewed as having an area between the area of the largest inscribed rectangle and the area of the smallest circumscribed rectangle, then, with the notation of the figure,

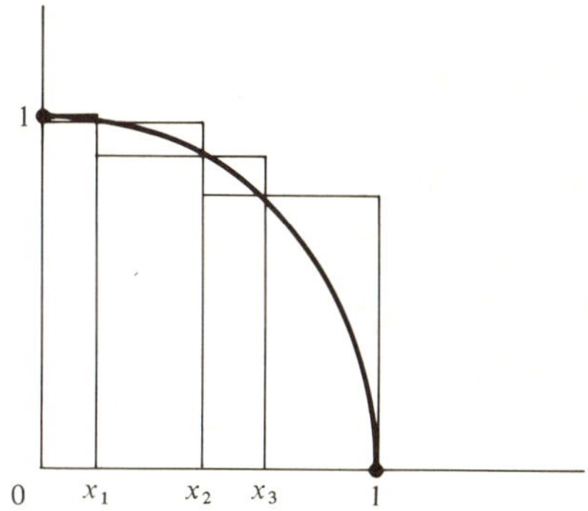

Figure 6.2

$$f(x_1)(x_1 - 0) + f(x_2)(x_2 - x_1) + f(x_3)(x_3 - x_2) + f(1)(1 - x_3) \leq A,$$
$$A \leq f(0)(x_1 - 0) + f(x_1)(x_2 - x_1) + f(x_2)(x_3 - x_2) + f(x_3)(1 - x_3).$$

Each term measures the area of a rectangle in Figure 6.2. The quarter circle can be cut up into vertical slices in many ways, but if this scheme is followed its area ought to lie between two such sums for every possible slicing. With some experimenting you may suspect that as the slicing becomes finer, the corresponding sums on the left (the lower sums) tend to increase, and those on the right (the upper sums) tend to decrease, squeezing A. (The trouble we will have with some proofs later will bolster any skepticism you may have, though.) Since we do not know A, we will verify the relation

$$\text{lub}[\text{set of all lower sums}] \leq \text{glb}[\text{set of all upper sums}]$$

and think of A being caught between these two numbers. If these numbers should be equal, then A is located.

There are many rather interesting functions for which graphs are practically impossible to draw. In case methods can be devised to study the sets of lower and upper sums, a number may be assigned the common value of the bounds of these sets, that is, to the function, which can reasonably be called the area under its graph.

For the physical example, take a spring attached to a wall with a bob on its free end. This is illustrated in Figure 6.3. Suppose that a force F is applied to the bob, directed along the axis of the spring so as to compress the spring. As the force compresses the spring, energy is stored in the spring as a result. We ask for a measure of this energy.

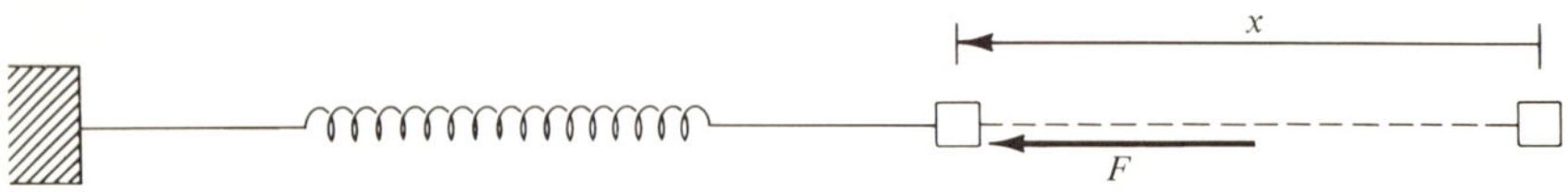

Figure 6.3

If zero force is applied to the bob when it is at rest, the bob stays at a fixed position relative to the wall, its rest position. When a force is applied to the bob against the spring, the bob moves to a new position, say a distance x, from the rest position. Suppose that initially the bob is moving through the rest position with constant positive velocity and that we adjust the force F so that this velocity stays constant; that is, no energy is being added to the kinetic energy of the bob, but the energy all goes into the spring. Hooke's law for springs says that under these conditions $F = kx$ for some constant k which depends on the units of measurement and the strength of the spring. It is sufficient for this illustration to assume that $k = 1$. Furthermore, the energy imparted to a system by a constant force is measured by the product of that force with the distance through which the force acts on the system. This is called the *work* done by a constant force on a system.

Clearly, the force F is not constant: it is 0 at the rest position; it is 1 at one unit distance from the rest position, and so on. Therefore, a direct application of the definition of work to this system is much too simple to be effective, and we resort to an approximating process. Let the bob move from position $x = a$ to position $x = b$, $0 < a < b$, and observe it at, say, six consecutive positions during this motion. See Figure 6.4.

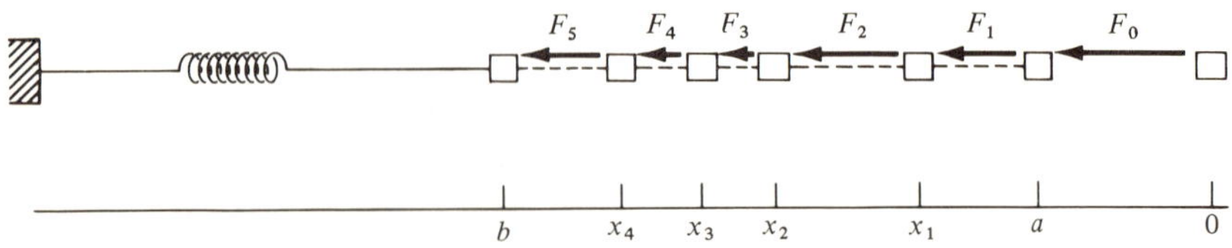

Figure 6.4

Between any two consecutive positions of the bob the force is least at the right-hand position and greatest at the left-hand position. If we approximate the work done on the spring by assuming that the force is constant in each step, and use the right-hand force for each one, we are surely underestimating the actual work W; if we use the left-hand force we are overestimating it. Thus,

$$a(x_1 - a) + x_1(x_2 - x_1) + x_2(x_3 - x_2) + x_3(x_4 - x_3) + x_4(b - x_4) \leqslant W,$$
$$W \leqslant x_1(x_1 - a) + x_2(x_2 - x_1) + x_3(x_3 - x_2) + x_4(x_4 - x_3) + b(b - x_4),$$

so we find a use for sums which are similar to those we used to approach the problem of the area of the circle. And again, if we contemplate all possible such time-lapse pictures of this motion, we are led to suppose that

$$\text{lub}[\text{set of all lower sums}] \leqslant W \leqslant \text{glb}[\text{set of all upper sums}].$$

If these two bounds are equal and we can compute them, then we have a reasonable value for W, which measures the energy imparted to the spring.

We now turn to a study of such upper and lower sums in a more general context.

6.1 PROBLEMS

1. Compute the two numbers

$$\sum_{j=1}^{j=10} x_{j-1}(x_j - x_{j-1}) \qquad \text{and} \qquad \sum_{j=1}^{j=10} x_j(x_j - x_{j-1})$$

for the sequence $x_0 = 0$, $x_1 = \frac{1}{2}$, $x_2 = \frac{2}{2}$, $x_3 = \frac{3}{2}$, . . . , $x_j = j/2$, . . . , $x_{10} = \frac{10}{2} = 5$.

2. (a) Draw the figure in the plane bounded below by the x-axis to the right by the line $x = 1$, and above by the graph of the function f, where $f(x) = 2x$.
   (b) Let $f(\underline{x}_j)$ be the minimum of f on $[x_{j-1}, x_j]$, and $f(\bar{x}_j)$ be the maximum of f on this interval. Compute the numbers

$$\sum_{j=1}^{j=4} f(\underline{x}_j)(x_j - x_{j-1}) \qquad \text{and} \qquad \sum_{j=1}^{j=4} f(\bar{x}_j)(x_j - x_{j-1})$$

   for the sequence $x_0 = 0$, $x_1 = \frac{1}{4}$, $x_2 = \frac{1}{2}$, $x_3 = \frac{3}{4}$, $x_4 = 1$.
   (c) Do the same for the sequence $x_0 = 0$, $x_1 = \frac{1}{10}$, $x_2 = \frac{3}{10}$, $x_3 = \frac{4}{10}$, $x_4 = \frac{7}{10}$, $x_5 = \frac{8}{10}$, $x_6 = 1$.
   (d) What do the numbers obtained in (b) and (c) tell us about the "area" of this figure?
   (e) Observe that this figure is a triangle, and compute its area. Compare with your answer to (d).
3. Carry out the program of Problem 2 for the figure bounded below by the x-axis, on the left by the y-axis, and above by the graph of g, where $g(x) = 3 - 3x$.

4. Consider the figure bounded below by the x-axis, on the right by the line $x = 2$, and above by the graph of h, where $h(x) = x^3$. Find upper and lower bounds for the "area" of this figure by using first the sequence $x_0 = 0$, $x_1 = 1$, $x_2 = 2$ and then the sequence $x_0 = 0$, $x_1 = \frac{1}{2}$, $x_2 = \frac{3}{2}$, $x_3 = 2$.
5. Consider the figure bounded below by the x-axis, on the left and right by the lines $x = -2$ and $x = 2$, and above by the graph of $\phi(x) = 1 + x^2$. Find upper and lower bounds for the "area" of this figure by using first the sequence $x_0 = -2$, $x_1 = 0$, $x_2 = 2$; next the sequence $x_0 = -2$, $x_1 = -1$, $x_2 = 0$, $x_3 = 1$, $x_4 = 2$; and finally the sequence $x_0 = -2$, $x_1 = -1$, $x_2 = 0$, $x_3 = 1$, $x_4 = \frac{3}{2}$, $x_5 = 2$.
6. Suppose that Hooke's law for a certain spring is $F = 4x$. Find upper and lower bounds for the energy imparted to the spring when it is compressed through the interval $x = 1$ to $x = 3$, by means of the sequence $a = 1$, $x_1 = \frac{3}{2}$, $x_2 = 2$, $x_3 = \frac{5}{2}$, $b = 3$.
7. Suppose that a certain spring is nonuniform, so that Hooke's law is replaced by $F = \sqrt{x}$, $x \geqslant 0$. Find two pairs of upper and lower bounds for the energy imparted to the spring when it is compressed through the interval $x = 0$ to $x = 16$, using each of the sequences

$$a = 0, \quad x_1 = 8, \quad b = 16$$
$$a = 0, \quad x_1 = 4, \quad x_2 = 9, \quad b = 16.$$

6.2 PRELIMINARY DEFINITIONS

Consider a function f whose domain is a closed interval $[a,b]$, and which is bounded. There is a number M such that

$$|f(x)| \leqslant M, \quad x \in [a,b].$$

Introduce a finite set of numbers π_n, ordered so that

$$\pi_n = \{x_0, x_1, \ldots, x_n | a = x_0 < x_1 < , \ldots, < x_{n-1} < x_n = b\}.$$

DEFINITION 6.1 The set π_n is an *n-partition* of the interval $[a,b]$.[1]

DEFINITION 6.2 The closed interval $[x_{j-1}, x_j]$ is the *jth subinterval* of the n-partition π_n.

Since f is bounded, the restriction of f to the jth subinterval is bounded; that is, the set

$$F_j = \{f(x) | x_{j-1} \leqslant x \leqslant x_j\}$$

[1] We will frequently drop the n and simply call π_n a *partition* of $[a,b]$, representing it by π when the exact number of subintervals is not crucial to the argument.

is bounded, and by the least upper bound axiom this set has the least upper bound $\bar{b}_j$ and the greatest lower bound $\underline{b}_j$,

$$\bar{b}_j = \text{lub } F_j \qquad \underline{b}_j = \text{glb } F_j.$$

Form the sums

$${}_a^b\bar{S}(\pi_n) = \sum_{j=1}^{n} \bar{b}_j(x_j - x_{j-1}) \qquad {}_a^b\underline{S}(\pi_n) = \sum_{j=1}^{n} \underline{b}_j(x_j - x_{j-1}).$$

In these sums

$$\bar{b}_j \geqslant -M, \quad \underline{b}_j \leqslant M, \quad j = 1,2,3, \ldots, n,$$

and therefore

$${}_a^b\bar{S}(\pi_n) \geqslant -M(b-a), \quad {}_a^b\underline{S}(\pi_n) \leqslant M(b-a)$$

for every π_n. Again, by the least upper bound axiom both of

$$\text{glb}\{{}_a^b\bar{S}(\pi_n) \mid \text{for all } \pi_n\}, \quad \text{lub}\{{}_a^b\underline{S}(\pi_n) \mid \text{for all } \pi_n\}$$

exist. We make the

DEFINITION 6.3 The *upper integral* of the function f on the interval $[a,b]$ is

$$\overline{\int_a^b} f = \text{glb}\{{}_a^b\bar{S}(\pi_n) \mid \text{all } \pi_n\}.$$

The *lower integral* of the function f on the interval $[a,b]$ is

$$\underline{\int_a^b} f = \text{lub}\{{}_a^b\underline{S}(\pi_n) \mid \text{all } \pi_n\}.$$

Let us examine what this rather complicated construction means.

EXAMPLE See Figure 6.5.

$$\begin{aligned} f(x) &= x\,, & 0 \leqslant x \leqslant 1, &\quad x \text{ rational} \\ &= -2x, & 0 < x < 1, &\quad x \text{ irrational} \end{aligned}$$

This function is bounded, $M = 2$. One partition of $[0,1]$ is

$$\pi_3 = \{0, \tfrac{1}{3}\sqrt{2}, \tfrac{3}{4}, 1\}.$$

Then

$$\begin{aligned} F_1 &= \text{rational } [0, \tfrac{1}{3}\sqrt{2}) \cup \text{irrational } [-\tfrac{2}{3}\sqrt{2}, 0) \\ F_2 &= \text{rational } (\tfrac{1}{3}\sqrt{2}, \tfrac{3}{4}] \cup \text{irrational } (-\tfrac{3}{2}, -\tfrac{2}{3}\sqrt{2}] \\ F_3 &= \text{rational } [\tfrac{3}{4}, 1] \cup \text{irrational } (-2, -\tfrac{3}{2}) \end{aligned}$$

and

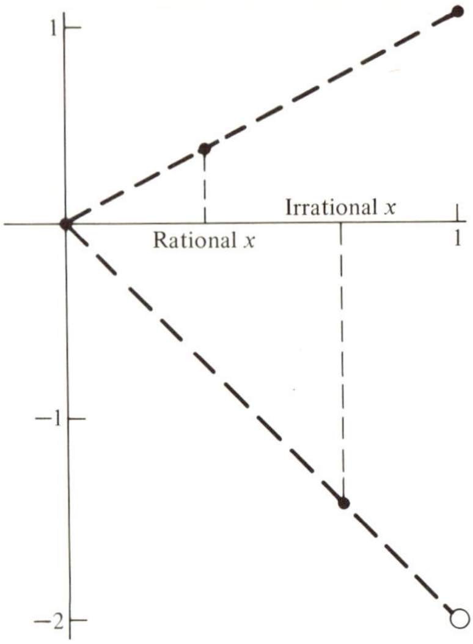

Figure 6.5

$$\begin{aligned} \bar{b}_1 &= \tfrac{1}{3}\sqrt{2} \notin F_1 & \underline{b}_1 &= -\tfrac{2}{3}\sqrt{2} \in F_1 \\ \bar{b}_2 &= \tfrac{3}{4} \in F_2 & \underline{b}_2 &= -\tfrac{3}{2} \notin F_2 \\ \bar{b}_3 &= 1 \in F_3 & \underline{b}_3 &= -2 \notin F_3. \end{aligned}$$

We form the sums

$$\begin{aligned} {}_0^1\bar{S}(\pi_3) &= \tfrac{1}{3}\sqrt{2}(\tfrac{1}{3}\sqrt{2} - 0) + \tfrac{3}{4}(\tfrac{3}{4} - \tfrac{1}{3}\sqrt{2}) \\ &\qquad + 1(1 - \tfrac{3}{4}) = \tfrac{149}{44} - \tfrac{1}{4}\sqrt{2} \cong 3.04 \\ {}_0^1\underline{S}(\pi_3) &= -\tfrac{2}{3}\sqrt{2}(\tfrac{1}{3}\sqrt{2} - 0) - \tfrac{3}{2}(\tfrac{3}{4} - \tfrac{1}{3}\sqrt{2}) \\ &\qquad - 2(1 - \tfrac{3}{4}) = -\tfrac{149}{72} + \tfrac{1}{2}\sqrt{2} \cong -1.36. \end{aligned}$$

There are many other partitions of type π_3, and of course many more of type π_n with more numbers in them, each yielding an upper and a lower sum. We have just proved, however, that the collection of all upper sums has a greatest lower bound, and the collection of all lower sums has a least upper bound, since f is a bounded function, and we proceed to find these upper and lower integrals of f.

The computation of the upper sums can be made for the general case as follows. Notice that on the jth subinterval

$$F_j = \{y | x_{j-1} \leqslant y \leqslant x_j, \quad y \text{ rational}\} \\ \cup \{y | -2x_j \leqslant y \leqslant -2x_{j-1}, \quad y \text{ irrational}\},$$

and therefore

$$\bar{b}_j = \text{lub } F_j = x_j,$$

so that each upper sum is

$$ {}_0^1\bar{S}(\pi_n) = \sum_{j=1}^{n} \bar{b}_j(x_j - x_{j-1}) = \sum_{j=1}^{n} x_j(x_j - x_{j-1}), $$

summed over the subintervals of the partition π_n. We seek the greatest lower bound of the set of these upper sums for all partitions of type π_n, for all n. By simple algebra,

$$ x_j(x_j - x_{j-1}) = \tfrac{1}{2}(x_j^2 - x_{j-1}^2) + \tfrac{1}{2}(x_j - x_{j-1})^2, $$

and so

$$ \text{(A)} \qquad {}_0^1\bar{S}(\pi_n) = \frac{1}{2}\sum_{j=1}^{n} (x_j^2 - x_{j-1}^2) + \frac{1}{2}\sum_{j=1}^{n} (x_j - x_{j-1})^2. $$

The first of the sums on the right side of line (A) is easily computed to be

$$ \begin{aligned} \sum_{j=1}^{n} (x_j^2 - x_{j-1}^2) &= \sum_{j=1}^{n} x_j^2 - \sum_{j=1}^{n} x_{j-1}^2 = [x_n^2 + \sum_{j=1}^{n-1} x_j^2] - [x_0^2 + \sum_{j=2}^{n} x_{j-1}^2] \\ &= x_n^2 - x_0^2 = 1, \quad (x_0 = 0,\ x_n = 1) \end{aligned} $$

for every π_n and all n.

For the second sum in line (A) think of a particular n, and remember that because we consider all partitions of type π_n, we have in particular the evenly spaced partition

$$ \pi_n = \{0, \frac{1}{n}, \frac{2}{n}, \ . \ . \ . \ , \frac{j-1}{n}, \frac{j}{n}, \ . \ . \ . \ , 1\} $$

for which this second sum is

$$ \sum_{j=1}^{n} \left(\frac{j}{n} - \frac{j-1}{n}\right)^2 = \frac{1}{n^2}\sum_{j=1}^{n} 1 = \frac{1}{n}. $$

Finally, then,

$$ {}_0^1\bar{S}(\pi_n) = \frac{1}{2}\left[1 + \sum_{j=1}^{n} (x_j - x_{j-1})^2\right] \geqslant \frac{1}{2} \quad \text{for all } \pi_n. $$

If ${}_0^1\bar{S}(\pi_m) = \frac{1}{2} + \epsilon$, for $\epsilon > 0$ and some π_m, we have only to take $n > (1/2\epsilon)$ to find a partition (the evenly spaced one) for which

$$ {}_0^1\bar{S}(\pi_n) = \frac{1}{2} + \frac{1}{2n} < \frac{1}{2} + \epsilon = {}_0^1\bar{S}(\pi_m), $$

and therefore

$$ \bar{\int_0^1} f = \text{glb}\{{}_0^1\bar{S}(\pi_n)\,|\text{all } \pi_n\} = \frac{1}{2}. $$

If you look at Figure 6.6, the graph of the rational part of f, you can

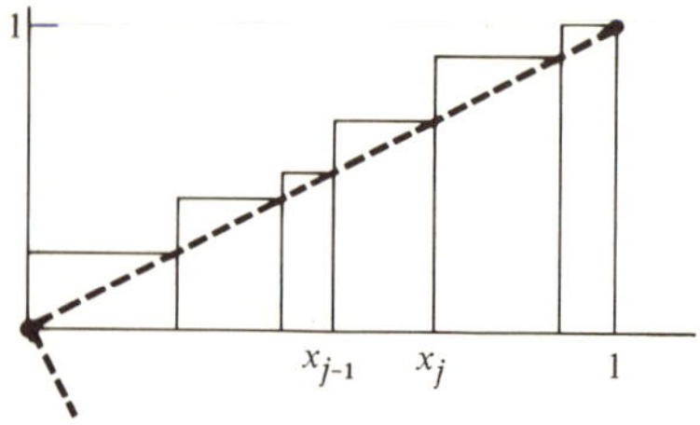

Figure 6.6

see the first term in line (A) is the area of the triangle below the dotted line and that the second term adds up the areas of the small triangles above the dotted line. It is questionable whether the graph of f really bounds an area, though, because its graph does not contain lines in the ordinary geometric sense. Therefore, we made algebra do the work that our definition of the upper integral calls for to compute it.

You should amuse yourself by computing

$$\underline{\int_0^1} f = -1,$$

different from the upper integral.

EXAMPLE $g(x) = x, \quad 0 \leqslant x \leqslant 1$

In this case the computation for

$$\overline{\int_0^1} g = \frac{1}{2}$$

is very similar to the computation of the upper integral of f in the previous example. Run through the justification for

$$\underline{\int_0^1} g = \frac{1}{2}.$$

The upper and lower integrals of g are equal. The graph of g, as shown in Figure 6.7, is a geometric line, and it does form with the axes a triangle that has area in the ordinary sense. Also notice that g is continuous, whereas f was not.

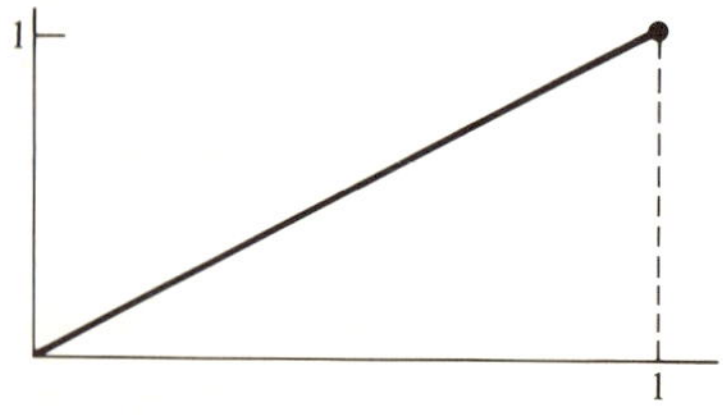

Figure 6.7

6.2 PROBLEMS

1. Prove that if

$$\begin{aligned} f(x) &= 0, \quad 0 \leqslant x \leqslant 1, \quad x \text{ rational} \\ &= 1, \quad 0 \leqslant x \leqslant 1, \quad x \text{ irrational}, \end{aligned}$$

then

$$\overline{\int_0^1} g = 1 \qquad \underline{\int_0^1} f = 0.$$

2. Prove that if

$$f(x) = 4x, \quad 0 \leqslant x \leqslant 1,$$

then

$$\overline{\int_0^1} f = \underline{\int_0^1} f = 2.$$

3. Let f be the function of Problem 15, page 53. What can you say about the upper and lower integrals of f?
4. Let

$$\begin{aligned} f(x) &= x, \quad x = 1 - \frac{1}{n}, \quad n = 1,2,3, \ldots, \\ &= 0, \quad \text{all other values of } x \in [0,1]. \end{aligned}$$

What can you say about the upper and lower integrals of f?
5. Let

$$f(x) = x^2, \quad 0 \leqslant x \leqslant 1.$$

(a) Consider

$$\pi_n = \left\{0, \frac{1}{n}, \frac{2}{n}, \ldots, \frac{j-1}{n}, \frac{j}{n}, \ldots, \frac{1}{n} \middle| \, n \text{ is a positive integer}\right\}.$$

Prove that

$$\text{glb}\{{}_0^1\bar{S}(\pi_n)\}$$

exists for these evenly spaced partitions.

(b) Prove that lub$\{{}_0^1\underline{S}(\pi_n)\}$ exists for these partitions.

(c) Show that the two numbers found in (a) and (b) are equal.

(d) Can you deduce from this that the upper and lower integrals of f exist and are equal?

6.3 THE INTEGRAL OF A CONTINUOUS FUNCTION ON A CLOSED INTERVAL

We have seen that a bounded function on a closed interval has an upper integral and a lower integral, and that these are not necessarily equal. However, the only functions we could discover for which these integrals are unequal were discontinuous functions. This section establishes that the upper and lower integrals of a continuous function are indeed equal. The domain of any function discussed in this section is understood to be a closed interval $[a,b]$.

The restriction of a continuous function f to a closed subinterval is also continuous. Therefore, the least upper bound of the set

$$F_j = \{f(x) | x_{j-1} \leq x \leq x_j\}$$

is the maximum value of f at some number $\bar{x}_j$ in the subinterval; that is,

$$\bar{b}_j = \text{lub } F_j = f(\bar{x}_j), \quad x_{j-1} \leq \bar{x}_j \leq x_j.$$

Similarly,

$$\underline{b}_j = \text{glb } F_j = f(\underline{x}_j), \quad x_{j-1} \leq \underline{x}_j \leq x_j.$$

Therefore, for continuous functions,

$$_a^b\bar{S}(\pi) = \sum_{j=1}^{n} f(\bar{x}_j)(x_j - x_{j-1}), \qquad _a^b\underline{S}(\pi) = \sum_{j=1}^{n} f(\underline{x}_j)(x_j - x_{j-1}),$$

where each $\bar{x}_j$ and $\underline{x}_j$ are in $[x_{j-1},x_j]$.

Lemma 1 to Theorem 6.1: Let f be a continuous function on $[a,b]$. Suppose that $_1\pi$ and $_2\pi$ are partitions of $[a,b]$, with $_1\pi \subset {}_2\pi$. Then

$$_a^b\bar{S}(_1\pi) \geq {}_a^b\bar{S}(_2\pi) \qquad \text{and} \qquad _a^b\underline{S}(_1\pi) \leq {}_a^b\underline{S}(_2\pi).$$

Proof: This lemma says that as a partition is refined, that is, more points are inserted, the upper sum tends to decrease and the lower sum tends to increase.

Each partition is a finite set of numbers. If we start with $_1\pi$, we can construct $_2\pi$ by inserting one number at a time into $_1\pi$ and arrive at $_2\pi$ in a finite number of steps.

If $_1\pi = {}_2\pi$, then clearly the equality holds in the formulas of the lemma.

If $_1\pi \neq {}_2\pi$, then there is a number $y_J \in {}_2\pi$, $y_J \notin {}_1\pi$. If $_1\pi$ is an n-partition, consider the $(n+1)$-partition $_1\pi_{n+1} = {}_1\pi_n \cup \{y_J\}$, $x_{J-1} < y_J < x_J$, where the x's $\in {}_1\pi_n \cap {}_2\pi$. Observe the sum

$$\begin{aligned}
_a^b\bar{S}(_1\pi_n) &= \sum_{j=1}^{n} f(\bar{x}_j)(x_j - x_{j-1}), \quad x_{J-1} < y_J < x_J \\
&= \sum_{j=1}^{J-1} f(\bar{x}_j)(x_j - x_{j-1}) + f(\bar{x}_J)(x_J - x_{J-1}) \\
&\quad + \sum_{j=J+1}^{n} f(\bar{x}_j)(x_j - x_{j-1}) \\
&= A + f(\bar{x}_J)(x_J - x_{J-1}) + B,
\end{aligned}$$

and note the equality

$$f(\bar{x}_J)(x_J - y_{J-1}) = f(\bar{x}_J)(y_J - x_{J-1}) + f(\bar{x}_J)(x_J - y_J),$$

where $x_{J-1} < \bar{x}_J < x_J$. Let $\bar{s}_J$ and $\bar{t}_J$ be the numbers at which

$$\begin{array}{lll} f(\bar{s}_J) & \text{is maximum on} & [x_{J-1},y_J] \\ f(\bar{t}_J) & \text{is maximum on} & [y_J,x_J]. \end{array}$$

Then, since $f(\bar{x}_J)$ is the maximum of f on $[x_{J-1},x_J]$, we have

$$f(\bar{s}_J) \leqslant f(\bar{x}_J) \qquad \text{and} \qquad f(\bar{t}_J) \leqslant f(\bar{x}_J),$$

and

$$\begin{aligned} {}_a^b\bar{S}({}_1\pi_n) &= A + f(\bar{x}_J)(y_J - x_{J-1}) + f(\bar{x}_J)(x_J - y_J) + B \\ &\geqslant A + f(\bar{s}_J)(y_J - x_{J-1}) + f(\bar{t}_J)(x_J - y_J) + B \\ &= {}_a^b\bar{S}({}_1\pi_{n+1}). \end{aligned}$$

Thus, the introduction of one new number into ${}_1\pi$ leads to a new sum which does not exceed ${}_a^b\bar{S}({}_1\pi)$. After a finite number of steps ${}_1\pi_{n+k}$ will equal ${}_2\pi$, and this proves the first inequality of the lemma.

Write out the proof of the second inequality for yourself. Notice that this result is independent of the positiveness or negativeness of the values of f on $[a,b]$. ■

Lemma 2 to Theorem 6.1: Let f be a continuous function on $[a,b]$. Then

$$\underline{\int_a^b} f \leqslant \overline{\int_a^b} f.$$

Proof: Since f is necessarily a bounded function, both integrals exist.

On any subinterval $[x_{j-1},x_j]$ of a partition π

$$f(\underline{x}_j) = \min f \leqslant \max f = f(\bar{x}_j)$$

and therefore

$$\text{(A)} \qquad {}_a^b\underline{S}(\pi) \leqslant {}_a^b\bar{S}(\pi).$$

Let ${}_1\pi$ and ${}_2\pi$ be any two partitions. Then ${}_1\pi \cup {}_2\pi$ is a partition having ${}_1\pi$ and ${}_2\pi$ as subsets. By Lemma 1 and line (A),

$${}_a^b\underline{S}({}_1\pi) \leqslant {}_a^b\underline{S}({}_1\pi \cup {}_2\pi) \leqslant {}_a^b\bar{S}({}_1\pi \cup {}_2\pi) \leqslant {}_a^b\bar{S}({}_2\pi),$$

that is, no lower sum ever exceeds any upper sum, whatever partitions are used, and any lower sum is a lower bound of the set of all upper sums. Hence, for all partitions π,

$${}_a^b\underline{S}(\pi) \leqslant \text{glb}\{{}_a^b\bar{S}(\pi)\} = \overline{\int_a^b} f.$$

The upper integral is, then, an upper bound of the set of all lower sums, and

$$\underline{\int_a^b} f = \text{lub}\{{}_a^b\underline{S}(\pi)\} \leq \overline{\int_a^b} f. \blacksquare$$

6.3 PROBLEMS

1. Let $f(x) = x, \quad 0 \leq x \leq 1$.
   (a) Let $\pi_n = \{0, 1/n, 2/n, \ldots, 1 | n \text{ is a positive integer}\}$, the evenly spaced partition. Compute each of the following.

   $${}_0^1\underline{S}(\pi_4), \qquad {}_0^1\overline{S}(\pi_4), \qquad {}_0^1\underline{S}(\pi_{10}), \qquad {}_0^1\overline{S}(\pi_{10}).$$

   (b) Show that

   $${}_0^1\underline{S}(\pi_n) = \frac{1}{2} - \frac{1}{2n} \qquad \text{and} \qquad {}_0^1\overline{S}(\pi_n) = \frac{1}{2} + \frac{1}{2n}.$$

   Hint: $\sum_{j=1}^{n} j = \frac{1}{2}n(n+1)$
   (c) Prove that $\text{lub}\{1/2 - (1/2n)\} = \frac{1}{2}$.
   (d) Prove that $\underline{\int_0^1} f \geq \frac{1}{2}$. Remember to consider partitions other than just the evenly spaced ones.
   (e) Prove that $\overline{\int_0^1} f \leq \frac{1}{2}$.
   (f) Prove that $\underline{\int_0^1} f = \overline{\int_0^1} f$.
   Hint: Lemma 2
2. Let $f(x) = x^2$, $0 \leq x \leq 1$, and prove that $\underline{\int_0^1} f = \overline{\int_0^1} f = \frac{1}{3}$.

   Hint: Follow the pattern of Problem 1 and use $\sum_{j=1}^{n} j^2 = \frac{1}{6}n(n+1)$ $(2n+1)$.
3. Let $f(x) = 2(1-x)$, $0 \leq x \leq 1$, and prove that $\underline{\int_0^1} f = \overline{\int_0^1} f = 1$.
4. Prove that, for a function f which is continuous on $[a,b]$,

   $$\underline{\int_a^b} f = -\overline{\int_a^b} (-f).$$

   Hint: On any subinterval the minimum value of f is the maximum value of $-f$. Apply the greatest lower bound form of the least upper bound axiom.

Lemma 3 to Theorem 6.1: Let f be a continuous function on $[a,b]$. Then there exist numbers m and M, the extreme values of f, such that

$$m(b-a) \leqslant \underline{\int_a^b} f \qquad \text{and} \qquad \overline{\int_a^b} f \leqslant M(b-a).$$

Proof: Since f is necessarily bounded on $[a,b]$, there exist numbers m and M the extreme values of f, such that

$$f(\underline{x}) = m \leqslant f(x) \leqslant M = f(\bar{x}), \quad x \in [a,b].$$

Consider the partition $\pi = \{a,b\}$. Then

$${}_a^b\bar{S}(\pi) = f(\bar{x})(b-a), \quad a \leqslant \bar{x} \leqslant b$$

is one of the upper sums in the collection of which $\overline{\int_a^b}$ is a lower bound. Therefore,

$$\overline{\int_a^b} f \leqslant {}_a^b\bar{S}(\pi) = f(\bar{x})(b-a) = M(b-a).$$

The other inequality arises similarly. ■

For both of the integrals $\overline{\int_a^b} f$ and $\underline{\int_a^b}$ we assumed that $a < b$. If $a \geqslant b$, the integral symbols have not been used. We bring them into our discussion by the next definition.

DEFINITION 6.4 If $a \in \operatorname{dom} f$, then $\underline{\int_a^a} f = \overline{\int_a^a} f = 0$. If $a > b$, and $\underline{\int_b^a} f$, $\overline{\int_b^a} f$ exist, then

$$\underline{\int_a^b} f = -\underline{\int_b^a} f \qquad \text{and} \qquad \overline{\int_a^b} f = -\overline{\int_b^a} f.$$

If you recall the way the lower and upper sums are built and the way the partitions are ordered, you can see that Definition 6.4 is natural.

Lemma 4 to Theorem 6.1: Let f be a continuous function on an interval I. If a, b, and c are numbers in I, then

$$\underline{\int_a^c} f + \underline{\int_c^b} f = \underline{\int_a^b} f \qquad \text{and} \qquad \overline{\int_a^c} f + \overline{\int_c^b} f = \overline{\int_a^b} f.$$

Proof: Suppose that $a < c < b$, and let ${}_1\pi$ be a partition of $[a,c]$ and ${}_2\pi$ be a partition of $[c,b]$, so that ${}_1\pi \cup {}_2\pi$ is a partition of $[a,b]$. Then

$${}_a^c\underline{S}({}_1\pi) + {}_c^b\underline{S}({}_2\pi) = {}_a^b\underline{S}({}_1\pi \cup {}_2\pi) \leqslant \underline{\int_a^b} f$$

since the lower integral is an upper bound of the set of lower sums. When we write this inequality as

$$ {}_a^c\underline{S}({}_1\pi) \leqslant \underline{\int}_a^b f - {}_c^b\underline{S}({}_2\pi), $$

we see that for any particular ${}_2\pi$, the right side is an upper bound for all lower sums on $[a,c]$, and therefore

$$ \underline{\int}_a^c f \leqslant \underline{\int}_a^b f - {}_c^b\underline{S}({}_2\pi). $$

From this

$$ {}_c^b\underline{S}({}_2\pi) \leqslant \underline{\int}_a^b f - \underline{\int}_a^c f $$

for any ${}_2\pi$, and as before

$$ \underline{\int}_c^b f \leqslant \underline{\int}_a^b f - \underline{\int}_a^c f \qquad \text{or} \qquad \underline{\int}_a^c f + \underline{\int}_c^b f \leqslant \underline{\int}_a^b f. $$

Next, let π_n be an n-partition of $[a,b]$, and $\pi_{n+1} = \pi_n \cup \{c\}$ in case $c \notin \pi_n$. In case $c \in \pi_n$, just replace π_{n+1} by π_n in the following argument. Take those numbers in $[a,c]$ which belong to π_{n+1} for ${}_1\pi$, and those numbers in $[c,b]$ which belong to π_{n+1} for ${}_2\pi$. Then $\pi_{n+1} = {}_1\pi \cup {}_2\pi$ and

$$ {}_a^c\underline{S}({}_1\pi) + {}_c^b\underline{S}({}_2\pi) = {}_a^b\underline{S}(\pi_{n+1}). $$

Since

$$ {}_a^c\underline{S}({}_1\pi) \leqslant \underline{\int}_a^c f \qquad \text{and} \qquad {}_c^b\underline{S}({}_2\pi) \leqslant \underline{\int}_c^b f, $$

then

$$ \underline{\int}_a^c f + \underline{\int}_c^b f \geqslant {}_a^b\underline{S}(\pi_{n+1}) \geqslant {}_a^b\underline{S}(\pi_n) $$

by Lemma 1 (π_{n+1} may have one more number than π_n has). Thus $\underline{\int}_a^c f + \underline{\int}_c^b f$ is an upper bound of all lower sums, and

$$ \underline{\int}_a^c f + \underline{\int}_c^b f \geqslant \underline{\int}_a^b f. $$

This proves the first equality for $a < c < b$.

In view of Definition 6.4 we can show that this formula is unaffected by the order of the numbers a, b, and c. Apparently one must examine, routinely, the cases. Two of these are

$$ c < a < b, \quad \underline{\int}_a^c f + \underline{\int}_c^b f = -\underline{\int}_c^a f + \underline{\int}_c^a f + \underline{\int}_a^b f = \underline{\int}_a^b f $$

and

$$a = c < b, \quad \underline{\int_a^c} f + \underline{\int_c^b} f = 0 + \underline{\int_c^b} f = \underline{\int_a^b} f.$$

When these cases are finished, the formula for the upper integrals is quickly implied by the observation of Problem 4, Section 6.3.

$$\overline{\int_a^c} f + \overline{\int_c^b} f = -\underline{\int_a^c} (-f) - \underline{\int_c^b} (-f) = -\left[\underline{\int_a^c} (-f) + \underline{\int_c^b} (-f)\right]$$
$$= -\underline{\int_a^b} (-f) = \overline{\int_a^b} f. \blacksquare$$

For the next lemma we must remember an important item in the definition of the integrals, both upper and lower. Let $x \in [a,b]$, where this interval is the domain of a bounded function f. Then the restriction of f to the interval $[a,x]$ is also bounded, and the number

$$\overline{\int_a^x} f = \text{glb}\{{}_a^x\bar{S}(\pi) \,|\, \pi \text{ is a partition of } [a,x]\}$$

is uniquely determined by x, since the greatest lower bound of a set is unique. Similarly, $\underline{\int_a^x} f$ is uniquely determined by x. The relations

$$\text{(A)} \qquad \bar{F}\colon \bar{F}(x) = \overline{\int_a^x} f \qquad \text{and} \qquad \underline{F}\colon \underline{F}(x) = \underline{\int_a^x} f, \quad x \in [a,b]$$

are functions, each with domain $[a,b]$. In view of Definition 6.4, if f is bounded on an interval I, and $a \in I$, the Formulas (A) define functions with domain I.

Lemma 5 to Theorem 6.1: Let f be a continuous function on an interval I, and $a \in I$. Then $\bar{F}$ and $\underline{F}$ are differentiable on I and

$$D\bar{F} = D\underline{F} = f$$

where $\bar{F}$ and $\underline{F}$ are defined in line (A).

Proof: The slope function for $\bar{F}$ at x is

$${}_{\bar{F}}\sigma_x(h) = \frac{1}{h}\,[\bar{F}(x+h) - \bar{F}(x)], \quad x \in I, \quad x+h \in I,$$
$$= \frac{1}{h}\left[\overline{\int_a^{x+h}} f - \overline{\int_a^x} f\right] = \frac{1}{h}\overline{\int_x^{x+h}} f$$

by Lemma 4. Now take $\epsilon > 0$; since f is a continuous function, there is a number $\delta > 0$ such that if $|h| < \delta$,

$$|f(x+h) - f(x)| < \frac{\epsilon}{2}.$$

Also if $|t| \leqslant |h|$, then

$$|f(x+t) - f(x)| < \frac{\epsilon}{2}$$

or

$$f(x) - \frac{\epsilon}{2} < f(x+t) < f(x) + \frac{\epsilon}{2}.$$

Combining Lemmas 2 and 3, then, we find

$$\left(f(x) - \frac{\epsilon}{2}\right) h \leqslant \bar{\int}_x^{x+h} f \leqslant \left(f(x) + \frac{\epsilon}{2}\right) h, \quad h > 0$$

and

$$\left(f(x) - \frac{\epsilon}{2}\right) (-h) \leqslant \bar{\int}_{x+h}^{x} f \leqslant \left(f(x) + \frac{\epsilon}{2}\right) (-h), \quad -h > 0.$$

Notice that Definition 6.4 gives

$$\frac{1}{-h} \bar{\int}_{x+h}^{x} f = \frac{1}{h} \bar{\int}_{x}^{x+h} f.$$

So either of the inequalities above implies

$$f(x) - \frac{\epsilon}{2} \leqslant \frac{1}{h} \bar{\int}_x^{x+h} f \leqslant f(x) + \frac{\epsilon}{2}$$

or

$$|{}_{\bar{F}}\sigma_x(h) - f(x)| \leqslant \frac{\epsilon}{2} < \epsilon, \quad |h| < \delta.$$

We have proved that

$$D\bar{F}(x) = \lim_{0} {}_{\bar{F}}\sigma_x = f(x), \quad x \in I,$$

and the proof for $D\underline{F} = f$ is nearly identical with this. ■

THEOREM 6.1 If f is a continuous function on the interval $[a,b]$, then

$$\bar{\int}_a^b f = \underline{\int}_a^b f.$$

Proof: Clearly, both these integrals exist. Let $\bar{F}$ and $\underline{F}$ be the functions defined by line (A) before Lemma 5. By Lemma 5

$$D\bar{F} = D\underline{F}$$

and the Corollary to Theorem 5.2 provides a constant function K so that

$$\bar{F} = \underline{F} + K.$$

Then

$$0 = \bar{F}(a) = \underline{F}(a) + K(a) = 0 + K(a) = K(a),$$

so K is the 0 constant function. Finally,

$$\overline{\int_a^b} f = \bar{F}(b) = \underline{F}(b) = \underline{\int_a^b} f. \blacksquare$$

DEFINITION 6.5 If f is a continuous function on $[a,b]$, the common value of the upper and lower integrals of f is the *integral* of f,

$$\int_a^b f.$$

Although we attend here chiefly to continuous functions, we do note that a discontinuous function may have an integral; see, for instance, Problem 4, Section 6.2. So we include the next definition.

DEFINITION 6.6 If dom $f = [a,b]$ and the upper and lower integrals exist and are equal, the function f is called *integrable* with the integral $\int_a^b f.$

The class of integrable functions contains the class of continuous functions on closed intervals as a proper subset but the integrable function class is larger.

6.3 PROBLEMS (*continued*)

5. Find $\int_a^b f$ for each of the following continuous functions. Do this by constructing the sets of upper and lower sums for evenly spaced partitions and applying Theorem 6.1.
   (a) $f(x) = 5, \quad a = 1, \quad b = 3$
   (b) $f(x) = 2x + 3, \quad a = 1, \quad b = 2$
   (c) $f(x) = 3x^2, \quad a = 0, \quad b = 1$
   (d) $f(x) = 3x^2, \quad a = -2, \quad b = 0$
   (e) $f(x) = 3x^2, \quad a = -2, \quad b = 1$
6. Use Definition 6.4 and Problem 5 to find $\int_a^b f$ for each of the following.
   (a) $f(x) = 5, \quad a = 3, \quad b = 1$
   (b) $f(x) = 3x^2, \quad a = 0, \quad b = -2$
   (c) $f(x) = \log \cos x, \quad a = \pi/7, \quad b = \pi/7$
   (d) $f(x) = 3x^2, \quad a = 1, \quad b = -2$

7. (a) Prove that $\int_{-4}^{4} f = 0$ when $f(x) = x^3$.
Hint: $f(-x) = -f(x)$
(b) Prove that

$$\int_{-a}^{a} f = 0 \text{ if } f(-x) = -f(x), \quad |x| \leqslant a.$$

8. Prove that if f is continuous on $[a,b]$, then the relation F,

$$F\colon F(x) = \int_{a}^{x} f, \quad x \in [a,b],$$

is a function, and that $DF = f$.

9. Deduce from Problem 8 that if F is continuously differentiable with $F(a) = 0$, then

$$\int_{a}^{x} DF = F(x).$$

10. Prove that if f is a continuous function, then

$$D\int_{a}^{x} f = f(x).$$

Hint: Compare this notation with Problem 8.

6.4 THE FUNDAMENTAL THEOREM OF INTEGRAL CALCULUS

THEOREM 6.2 (Fundamental theorem) Let f be a continuous function on an interval I, and let f be the derivative of a function $G, f = DG$. Then

$$\int_{a}^{b} f = G(b) - G(a), \quad a,b \in I.$$

Proof: The integral exists by Theorem 6.1. Also the restriction of f to $[a,x]$ is continuous, and therefore (see line (A) before Lemma 5 to Theorem 6.1) the function F,

$$F(x) = \bar{F}(x) = \bar{\int_{a}^{x}} f = \int_{a}^{x} f, \quad x \in I,$$

exists and, by Lemma 5,

$$DF = D\bar{F} = f = DG.$$

Therefore, $G = F + K$ for some constant function K, and

$$G(b) - G(a) = (F(b) - K(b)) - (F(a) - K(a)) = F(b) - F(a)$$
$$= \int_{a}^{b} f. \blacksquare$$

Corollary to Theorem 6.2: (Linearity of the integral) Let f and g be continuous on $[a,b]$. Then

$$\int_a^b (f+g) = \int_a^b f + \int_a^b g,$$

and

$$\int_a^b kf = k\int_a^b f, \quad k \text{ a real number.}[2]$$

Proof: Set

$$F(x) = \int_a^x f, \quad G(x) = \int_a^x g$$

so that

$$DF = f, \quad DG = g, \quad D(F+G) = DF + DG = f + g, \quad D(kF) = kf.$$

Then

$$\begin{aligned}\int_a^b (f+g) &= (F+G)(b) - (F+G)(a)\\ &= (F(b) - F(a)) + (G(b) - G(a))\\ &= \int_a^b f + \int_a^b g\end{aligned}$$

and

$$\int_a^b kf = kF(b) - kF(a) = k(F(b) - F(a)) = k\int_a^b f. \blacksquare$$

Certain conventions about the notation with which an integral is written are suggested by the following three examples.

EXAMPLE 1 Find $\int_2^4 f$, where $f(x) = 3x^2$. We know that if $G(x) = x^3$, then $DG = f$. Therefore, by the fundamental theorem

$$\int_2^4 f = 4^3 - 2^3 = 56.$$

EXAMPLE 2 Let $F(y) = \int_1^2 {}_yf$, where ${}_yf(x) = 12y^2x^3$. Find a formula for $F(y)$. For each y there is a ${}_yG$; that is, ${}_yG(x) = 3y^2x^4$, such that

$$D_yG = {}_yf.$$

[2] This proposition is true under the weaker hypothesis that f and g be integrable. However, Theorem 6.2 is not then available for the proof, which becomes harder.

Therefore,

$$F(y) = \int_1^2 {}_yf = {}_yG(2) - {}_yG(1) = 48y^2 - 3y^2 = 45y^2.$$

EXAMPLE 3 Let $F(x) = \int_1^2 {}_xf$, where ${}_xf(y) = 12x^3y^2$. Find a formula for $F(x)$. For each x there is an ${}_xG$; that is, ${}_xG(y) = 4x^3y^3$, such that

$$D\,{}_xG = {}_xf.$$

Therefore,

$$F(x) = \int_1^2 {}_xf = {}_xG(2) - {}_xG(1) = 32x^3 - 4x^3 = 28x^3.$$

We have found it preferable on earlier occasions to use a formula for a function as a name for it. We have said, "Find the derivative of $x^2 - 4x$," meaning, "Find the derivative of the polynomial function whose formula is $x^2 - 4x$." And we have written $D(x^2 - 4x) = 2x - 4$ as a shorthand.

In Example 1 above we could say, "Find $\int_2^4 3x^2$," and find the number 56 mentally. For several reasons this is not done in the literature. One of the reasons is that the Examples 2 and 3 occur frequently in various guises. Suppose that we said, "Find $\int_1^2 12y^2x^3$." Do we mean that for each y we are asked to integrate a cubic polynomial function (Example 2)? Or do we mean that for each x we are to integrate a quadratic polynomial function (Example 3)?

A standard convention, when using a formula for the name of a function, is to indicate the symbol that is to stand for a number in the domain of the implied function with the prefix d (for "domain" if you wish). Examples 2 and 3 can be presented unambiguously as

EXAMPLE 2 Find $\int_1^2 y^2x^3dx$.

EXAMPLE 3 Find $\int_1^2 x^3y^2dy$.

There are other ways in which the d prefix is put to use, and we saw some of them in Section 4.3.7. Although Example 1 involves no ambiguity, it, too, is usually written,

EXAMPLE 1 Find $\int_2^4 3x^2 dx$.

6.4 PROBLEMS

1. Apply the fundamental theorem to find $\int_1^3 f$ when $f(x) = 5x$.
   (a) Take $G(x) = \frac{5}{2}x^2$.
   (b) Take $G(x) = \frac{5}{2}x^2 + 7$.
   (c) Take $G(x) = \frac{5}{2}x^2 - 3$.
   (d) Take $G(x) = \frac{5}{2}x^2 + \sin \pi/5$.

2. Find each of the following.
   (a) $\int_1^2 f,\ \ f(x) = 2x + 3$

   (b) $\int_2^{10} f,\ \ f(x) = x(x + 1)$

   (c) $\int_1^2 f,\ \ f(x) = x^3$

   (d) $\int_2^1 f,\ \ f(x) = x^3$

3. Write each of the integrals in Problem 2 in d prefix notation. For (a) this is $\int_1^2 (2x + 3)dx$.

4. Find each of the following.
   (a) $\int_9^{10} x^3\,dx$ (b) $\int_0^1 10x^3\,dx$

   (c) $\int_{-1}^0 x^3\,dx$ (d) $\int_{-1}^1 s^3\,ds$

   (e) $\int_1^3 (1 - x + x^2)dx$ (f) $\int_0^5 x^{10}\,dx$

   (g) $\int_0^4 \sqrt{b}\,db$ (h) $\int_1^2 1/\sqrt{x}\,dx$

   (i) $\int_1^2 1/w^2\,dw$ (j) $\int_{10}^{11} 1/x^2\,dx$

   (k) $\int_{1/2}^{3/2} (y^3 + y^{-3})dy$

5. Find each of the following.
   (a) $\int_{-3}^{-5} (x - x^2)dx$

(b) $\int_{-3}^{-5} (y - y^2)\,dy$

(c) $\int_{-3}^{-5} (w - w^2)\,dw$

6. Prove that $\int_a^b f(t)dt = \int_a^b f(w)dw$.

7. Prove that $\int_a^b f(s)ds + \int_b^a f(y)dy = 0$.

8. Find each of the following functions, and describe its domain.

   (a) $\int_3^{-2} xy^2\,dx$ (b) $\int_3^{-2} xy^2\,dy$

   (c) $\int_{-1}^{1} \frac{x}{y}\,dx$ (d) $\int_{-1}^{1} \frac{x}{y}\,dy$

   (e) $\int_2^5 a \sin k\,da$ (f) $\int_2^5 a \sin k\,dk$

9. Let

$$\begin{aligned} f(x) &= x + 1, \quad -1 \leqslant x \leqslant 0 \\ &= 1 - x, \quad 0 \leqslant x \leqslant 1. \end{aligned}$$

   Find $\int_{-1}^{1} f$ in two ways.

   (a) Show that f is continuous, so Lemma 4 to Theorem 6.1 is true for the integral. Take $a = -1$, $c = 0$, and $b = 1$, and apply the fundamental theorem to each of two integrals.

   (b) Prove that

$$\begin{aligned} G(x) &= \tfrac{1}{2}x^2 + x, \quad -1 \leqslant x \leqslant 0 \\ &= x - \tfrac{1}{2}x^2, \quad 0 \leqslant x \leqslant 1 \end{aligned}$$

   is differentiable, and $DG = f$. Apply the fundamental theorem to the given integral.

10. Prove that

$$f(x) = \frac{|x|}{x}, \quad 0 < |x| \leqslant 1$$
$$f(0) = 0$$

   is not the derivative of any function.

   Hints: If $DG = f$ on $[-1,1]$, then

   (a) $G(x) = -x + k, \quad x < 0$
   $\quad\quad = x + c \;, \quad x > 0.$

(b) Since G is differentiable, G is continuous; therefore $c = k$, and $G(x) = |x| + k$ for some k and $|x| \leqslant 1$.

(c) G has no derivative at 0.

11. Show that $\int_{-1}^{1} f = 0$ for the function f of Problem 10.

Hint: Form the upper and lower sums, with particular attention to each subinterval that contains 0. (If we used the function G of Problem 10, then $G(1) - G(-1) = (1 + k) - (1 + k) = 0$. Why is this application of the fundamental theorem not legitimate?)

12. Prove that if f is continuous on $[a,b]$, except at a finite number of numbers in $[a,b]$, and is bounded, then f is integrable on $[a,b]$.

6.5 APPLICATIONS OF THE INTEGRAL

Whenever the measure of a geometrical or physical quantity can be bounded between lower and upper sums of the types ${}_a^b\underline{S}$ and ${}_a^b\bar{S}$, respectively, it is worth searching for a function f so that $\int_a^b f$ measures that quantity. This technique is so strong that when it is not known in advance whether such a measure exists, if the tentative assumption of its existence leads to an integral, that integral is taken to define the measure of the quantity.

For the spring, since Hooke's law $F = kx$ can be described by the continuous function f, where

$$f(x) = kx, \quad a \leqslant x \leqslant b$$

we now know that $\int_a^b f$ exists, so this number is available and is used as a value of the work done to measure the energy imparted to the spring.

We now study some problems in which the integral can be used.

EXAMPLE A cylindrical tank 10 feet high has a radius of 5 feet and is full of a liquid with a density of 100 pounds per cubic foot. How much energy does it take to pump out the liquid over the rim?

This energy is measured by the work done in moving the liquid just to the top of the tank and letting it continuously flow away.

Imagine, as in Figure 6.8, a vertical scale up the side of the tank so that the height, x feet, of a plane section above the bottom is in the interval $[0,10]$. Let

$$\pi_n = \{0,x_1, \ . \ . \ . \ ,x_{j-1},x_j, \ . \ . \ . \ ,10\}$$

be a partition of $[0,10]$. Then the volume of liquid between two horizontal slices of the tank at levels x_{j-1} and x_j is

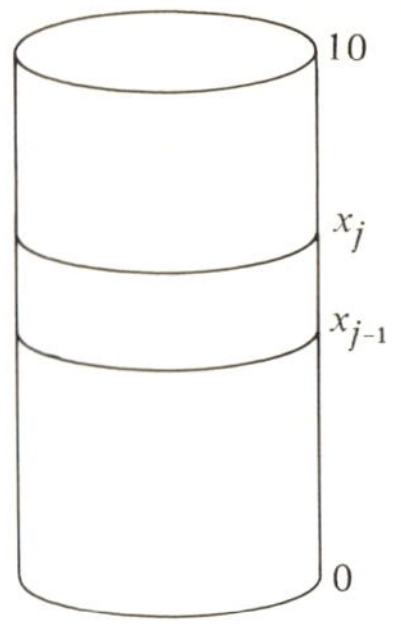

Figure 6.8

$$25\pi(x_j - x_{j-1})$$

and the force necessary to hold it there is its weight

$$2500\pi(x_j - x_{j-1}).$$

Suppose that we ignore the force necessary to accelerate the liquid from rest, and assume that this disk of liquid is moved slowly by a force equal to its weight through the distance to the rim. Consider two extreme possibilities for getting this slice of the liquid to the top of the tank. If we raise it until the top of the disk is even with the rim, it will have moved through a distance of $10 - x_j$, and the work done would be

$$2500\pi(x_j - x_{j-1})(10 - x_j).$$

If we could somehow freeze the liquid and raise the whole disk until its bottom is even with the rim, then the work done would be

$$2500\pi(x_j - x_{j-1})(10 - x_{j-1}).$$

Since, in fact, the liquid flows out as the disk passes the rim, the first estimate of the work is low and the second is high. Considering all the disks of liquid for the partition π_n, the work W done is bounded as

$$\sum_{j=1}^{n} 2500\pi(10 - x_j)(x_j - x_{j-1}) \leq W \leq \sum_{j=1}^{n} 2500\pi(10 - x_{j-1})(x_j - x_{j-1})$$

and "the cat is out of the bag." The ends of this inequality are recognizable as lower and upper sums ${}^{10}_{0}\underline{S}(\pi_n)$ and ${}^{10}_{0}\bar{S}(\pi_n)$ for the continuous function f,

$$f(x) = 2500\pi(10 - x), \quad 0 \leq x \leq 10,$$

and we know $\int_0^{10} f$ exists and is between every pair of approximating sums, and pairs of these sums exist arbitrarily close together. By the definition of this integral we take

$$W = \int_0^{10} f = \int_0^{10} 2500\pi(10-x)\,dx = 125{,}000\pi \text{ ft lb.}$$

The unit, ft lb, is carried over from the simple notion of 1 pound force acting through 1 foot, and is attached to this number to identify it as a value of the work done. This measures the energy used.

EXAMPLE If a constant force acts on a particle over a time interval, the product of the force with the time it is applied is called the *impulse I* applied to the particle. This impulse measures the momentum acquired by the particle as a result.

Suppose that the force F at time t is given by the continuous function f:

$$F = f(t), \quad \alpha \leqslant t \leqslant \beta.$$

Let π_n be a partition of $[\alpha,\beta]$ so that on the jth subinterval the impulse I_j is in between

$$f(\underline{t}_j)(t_j - t_{j-1}) \leqslant I_j \leqslant f(\bar{t}_j)(t_j - t_{j-1})$$

and therefore the total impulse I is between the sums,

$$\sum_{j=1}^{n} f(\underline{t}_j)(t_j - t_{j-1}) \leqslant I \leqslant \sum_{j=1}^{n} f(\bar{t}_j)(t_j - t_{j-1}).$$

Again, these are just the lower and upper sums for the continuous function f, which we know has the integral $\int_\alpha^\beta f$ between any such pair of sums, and, for fine enough partition π_n, these sums are arbitrarily close together. Notice that if the force were constant,

$$F = f(t) = c, \quad \alpha \leqslant t \leqslant \beta,$$

then

$$I = \int_\alpha^\beta f = c(\beta - \alpha),$$

recapturing the primitive definition of the impulse for a constant force.

Returning to the spring of Section 6.1, suppose that the bob is held at a point for which $x = a > 0$, so that the spring is compressed and then released. How do we measure the momentum imparted to the bob as it passes the rest position? In view of the discussion above we need a continuous function f so that

$$F = f(t), \quad \alpha \leqslant t \leqslant \beta, \quad f(\alpha) = a, \quad f(\beta) = 0.$$

Then $\int_\alpha^\beta f$ will be the answer. Take Hooke's law in the form $F = -x$, with $k = -1$ negative since the force on the bob is directed toward the rest position. (See Figure 6.9.)

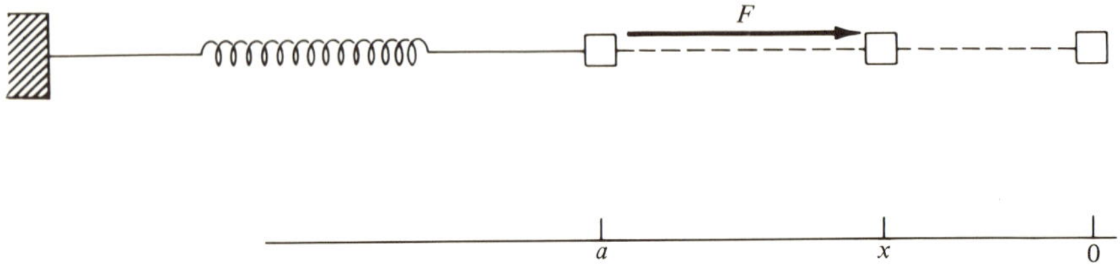

Figure 6.9

It would seem that if the bob is heavy it would take longer for the spring to push it back to its rest position than if it were light; that is, f should depend on the mass m of the bob. Furthermore, the force the spring can exert depends only on the distance x of the bob from the rest position; so we look for a relation s between x and t:

$$x = s(t), \quad \alpha \leqslant t \leqslant \beta, \quad s(\alpha) = a, \quad s(\beta) = 0.$$

From Hooke's law, $F = -x = -s(t)$, and by Newton's law

$$-s(t) = mD^2s(t).$$

That is, $D^2s = -(1/m)s$, so we are looking for a function that is, essentially, its own second derivative. The sine and the cosine are candidates, and in fact the function

$$S(t) = A \cos \frac{1}{\sqrt{m}} t + B \sin \frac{1}{\sqrt{m}} t \quad \text{for some } A, B$$

satisfies Newton's law for the problem. Since the bob starts from rest at $t = 0$

$$s(0) = a, \quad Ds(0) = 0$$

so that $A = a$, $B = 0$, and $s(t) = a \cos (1/\sqrt{m})t$. Thus, the function f can be defined by

$$f(t) = -a \cos \frac{1}{\sqrt{m}} t, \quad 0 \leqslant t \leqslant \frac{\pi}{2} \sqrt{m}$$

and the impulse I is given by

$$I = \int_0^{\pi/2\sqrt{m}} f = -a \int_0^{\pi/2\sqrt{m}} \cos \frac{1}{\sqrt{m}} t \, dt = -a \sqrt{m}.$$

This impulse is negative because we chose the coordinates so that the force acts in the negative direction. If the mass m of the bob is large, so is the impulse, and therefore the amount of momentum imparted, since the domain of f is a large interval.

6.5 PROBLEMS

1. When a certain spring has been stretched 1 foot, a force of 6 pounds is needed to hold it. How much work was done in stretching it this amount?
   Hint: We assume that Hooke's law is satisfied, so $f(x) = kx$. But $f(1) = 6$, so $k = 6$. Now follow the analysis of the spring example in Section 6.1 to show that
   $$W = \int_0^1 f,$$
   where
   $$f(x) = 6x.$$
   Answer: 3 ft lbs
2. How much work is done if the spring in Problem 1 is stretched another foot?
3. A spring hung vertically is found to be stretched from its natural length of 12 inches to 15 inches by a weight of 18 pounds. How much work is done in stretching the spring
   (a) from 12 to 15 inches?
   (b) from 13 to 16 inches?
   (c) from 14 to 17 inches?
4. To compress a spring from its natural length of 6 feet to a length of 5 feet, 25 ft lbs of work are needed. How much work is needed to compress it another foot?
   Answer: 75 ft lbs
5. A piece of wire, hanging down from a windlass, weighs 8 ounces per foot and is 200 feet long. How much work does it take to wind it up?
   Answer: 10,000 ft lbs
6. How much work is done in winding up the first half of the wire of Problem 5?
7. To the end of the wire of Problem 5 there is attached a bucket weighing 10 pounds containing 90 pounds of water. How much work is done in winding it up to the windlass?
8. In Problem 7 suppose that the wire is wound up at the rate of 50 feet per minute and suppose that the water leaks out at the rate of 2

pounds per minute. How much work is done in winding the bucket up to the windlass?

9. In Problem 8 suppose that the wire is wound up at the rate of 5 feet per minute. How much work is done in this case?
10. The force due to gravity acting on a particle of mass m at a distance r from the center of the earth is

$$F = \frac{mgR^2}{r^2},$$

where $g = 32$ (approximately) and R is the radius of the earth. How much work is done in raising a particle from the surface of the earth to a point 100 miles up? (Take $R = 4000$.) How much work does it take to raise it another 100 miles?
11. If the temperature of a gas remains constant, then its pressure and volume are connected by the equation

$$PV = k.$$

Suppose that the volume of a cylinder is 100 cubic feet, the area of the piston is 10 square feet, and the pressure is 1 pound per square foot. How much work is done in compressing the gas to 25 cubic feet? Can you evaluate the resulting integral?
12. If the pressure and volume of the gas of the previous problem are connected by the formula

$$PV^{3/2} = k,$$

find the work done.
13. Returning to the example, Section 6.5, in which the tank was pumped out, suppose that the liquid is pumped to a level 20 feet above the top of the tank. How much work is done?
Answer: $625{,}000\pi$ ft lbs
14. Suppose that the tank of Problem 13 were only half full to start with. How much work would be needed to pump it out?
15. A cylindrical tank is 10 feet high. Its base is an equilateral triangle, with each side 2 feet long. It is filled with a liquid weighing 75 pounds per cubic foot. How much work can be done by allowing the liquid to run out through a hole in the bottom of the tank?
Answer: $3750\sqrt{3}$ ft lbs
16. A cubical tank, whose volume is 1,000 cubic feet, is placed with its top 15 feet below the base of another cubical tank of the same dimensions. The lower tank is full of water. How much work does it take to pump the water into the upper tank if the connecting pipe runs to the top of the upper tank?
17. How much work is done if, in Problem 16, the connecting pipe runs to the base of the upper tank?

It is already clear that since the graph of a positive valued function f,

$$f(x) > 0, \quad a \leqslant x \leqslant b,$$

appears to be part of the boundary of a region enclosed by vertical lines through the points a,b on the x-axis, the interval $[a,b]$, and the graph of f, then the area of this region ought to be $\int_a^b f$. But, as we observed in the introduction, we have not produced a definition of the area of a figure with a curved edge, so the above comment is really circular. We need the following definition.

DEFINITION 6.7 If f is a continuous function on $[a,b]$ and $f(x) \geqslant 0$, $x \in \operatorname{dom} f$, the *area under the graph* of f is $\int_a^b f$.

EXAMPLE Suppose that $f(x) = 4 - x^2$, $-2 \leqslant x \leqslant 2$; then the area under the graph of f is $\int_{-2}^{2} f = \int_{-2}^{2} (4 - x^2)dx = 32/3$.

If the Definition 6.7 is a reasonable one, it ought to recapture the known measures of areas that have been geometrically defined, such as the area of a trapezoid.

EXAMPLE Let f be a linear function on $[a,b]$, for which $f(a) \geqslant 0$, $f(b) \geqslant 0$. Then (you should prove)

$$f(x) = \frac{f(b) - f(a)}{b - a}\, x + \frac{bf(a) - af(b)}{b - a}, \quad a \leqslant x \leqslant b, \quad f(x) \geqslant 0.$$

The area under the graph is represented by the trapezoid shown in Figure 6.10, which, by the geometry of polygons, has the area

$$\frac{1}{2}(b - a)[f(a) + f(b)].$$

But by Definition 6.7 the area under the graph of f is, by computation of

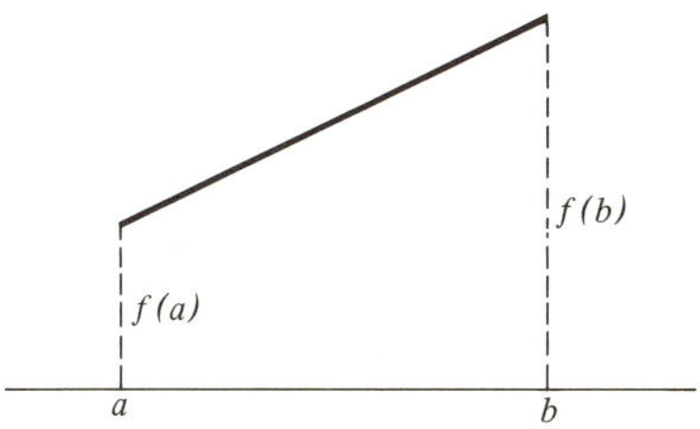

Figure 6.10

of the integral,

$$\int_a^b f = \frac{1}{2}(b-a)[f(a)+f(b)]$$

also.

EXAMPLE Let f be

$$f(x) = \sqrt{r^2 - x^2}, \quad 0 \leq x \leq 1, \quad r > 0$$

where $f(x) \geq 0$. With the coordinate axes, this graph (Figure 6.11) forms a quarter circle. The area under the graph of f, by Definition 6.7, is $\int_0^r f$. If you have done many differentiation problems, you may have run across the function G,

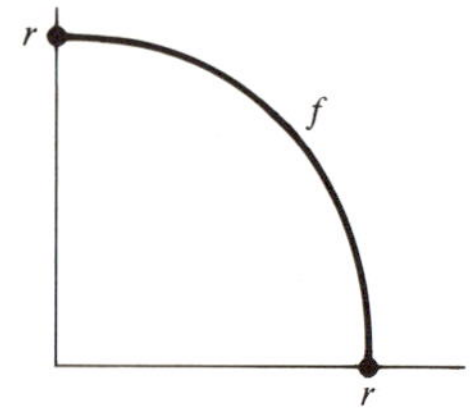

Figure 6.11

$$G(x) = \frac{r^2}{2} \arcsin \frac{x}{r} + \frac{1}{2} x\sqrt{r^2 - x^2}, \quad 0 \leq x \leq r,$$

and found that $DG = f$. (In the next chapter we will show how this function G can be discovered.) By the fundamental theorem

$$\int_0^r f = \int_0^r \sqrt{r^2 - x^2}\, dx = G(r) - G(0) = \frac{1}{4}\pi r^2.$$

When we define the area of the circle of radius r to be four times the area under the graph of this quarter circle, we arrive at the familiar formula πr^2 for the area of the circle.

Usually we equate this "area" of a plane geometric figure with the sum of the areas under the graphs of related functions.

EXAMPLE Find the area of the shaded part of Figure 6.12 where the the curve is the graph of f,

$$f(x) = x^2 - 4, \quad -3 \leq x \leq 2.$$

Consider the two shaded pieces separately.

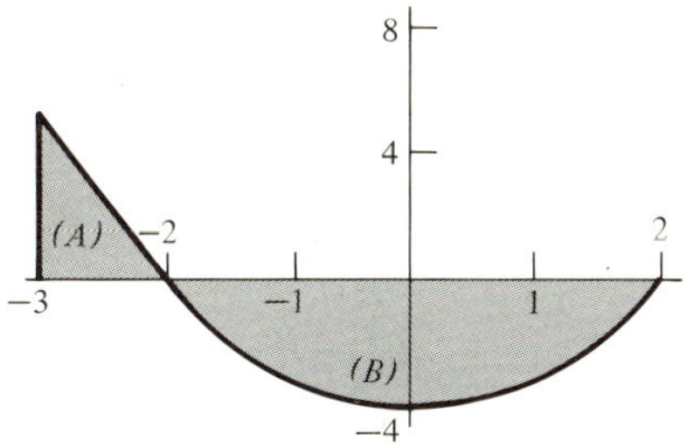

Figure 6.12

(A) $$\int_{-3}^{-2} f = \int_{-3}^{-2} (x^2 - 4)\,dx = \frac{7}{3}$$

Since the restriction of f to $[-2,2]$ takes only negative values $\int_{-2}^{2} f \leq 0$ (see Lemma 3 to Theorem 6.1, applied to a continuous function), so we take $\left|\int_{-2}^{2} f\right|$ for the area of the shaded portion (B).

(B) $$\left|\int_{-2}^{2} f\right| = \left|\int_{-2}^{2} (x^2 - 4)\,dx\right| = \left|-\frac{32}{3}\right| = \frac{32}{3}$$

The total shaded area is $7/3 + 32/3 = 13$. Notice that

$$\int_{-3}^{2} f = \int_{-3}^{2} (x^2 - 4)\,dx = -\frac{25}{3} = \frac{7}{3} - \frac{32}{3}.$$

EXAMPLE Find the area of the shaded part of Figure 6.13, where the curve is the graph of sin and the other boundary is a line. We consider the shaded area to be the area under the curve diminished by the area under the line. If f is

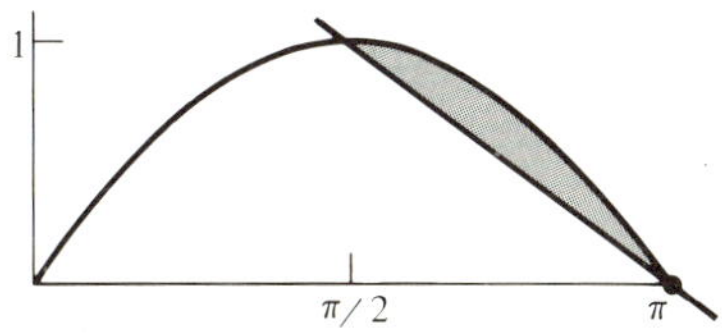

Figure 6.13

$$f(x) = \sin x, \quad \frac{\pi}{2} \leq x \leq \pi,$$

then

$$\int_{\pi/2}^{\pi} f = \int_{\pi/2}^{\pi} \sin x\,dx = 1.$$

If g is

$$g(x) = 2 - \frac{2}{\pi} x, \quad \frac{\pi}{2} \leqslant x \leqslant \pi,$$

then

$$\int_{\pi/2}^{\pi} g = \int_{\pi/2}^{\pi} \left(2 - \frac{2}{\pi} x\right) dx = \frac{\pi}{4}.$$

Therefore, the shaded area is $1 - \pi/4$.

By the Corollary to Theorem 6.2,

$$1 - \frac{\pi}{4} = \int_{\pi/2}^{\pi} f - \int_{\pi/2}^{\pi} g = \int_{\pi/2}^{\pi} (f - g),$$

and we could have computed the shaded area with a single integral.

6.5 PROBLEMS (*continued*)

18. Find the areas of the figures bounded below by the x-axis and laterally and above by the given lines and the graphs of the given functions.
   (a) $x = 1, x = 3;\ f(x) = x^3$
   Answer: 20
   (b) $x = 2, x = 4;\ f(x) = x^3$
   Answer: 60
   (c) $x = 0, x = 2;\ f(x) = x^3$
   (d) $x = a, x = b;\ f(x) = x^3$
   (e) $x = 1, x = 10;\ f(x) = x^4 + x^2$
   Answer: 20,332.8
   (f) $x = 1, x = 2;\ f(x) = 5x^2$
   (g) $x = 0, x = 1;\ f(x) = 10x^9$
   (h) $x = 0, x = 1;\ f(x) = 1 + x + x^2 + x^3 + x^4$
   (i) $x = 5, x = 10;\ f(x) = \sqrt{x}$
   Answer: $\dfrac{20\sqrt{10}}{3} - \dfrac{10\sqrt{5}}{3}$
   (j) $x = 1, x = 2;\ f(x) = x^4 - \sqrt[4]{x}$
   (k) $x = 1, x = 2;\ f(x) = 1/x^2$
   (l) $x = 2, x = 3;\ f(x) = 3x^3 - 3/x^3$
19. Find the area of the figure bounded above by the parabola $x^2 = 3 - y$ and below by the parabola $x^2 = y - 1$.
20. Find the areas of the figures bounded by the following pairs of curves.
   (a) $y = x$ and $y = 6x - x^2$
   Answer: 125/6
   (b) $y = x^3$ and $y = 2x$

(c) $y = x^4$ and $y = \sqrt{x}$
Answer: 7/15
(d) $y = 16x^2$ and $y = 4x^4$
(e) $y = (x - 3)^2$ and $y = 2x$

21. Find the area of the figure bounded above by the x-axis and below by the parabola $x^2 = y + 1$.
22. Find the areas of the figures bounded above by the x-axis and laterally and below by the graphs of the given equations.
(a) $x = 0$, $x = 1$; $y = -x^4$
Answer: $\frac{1}{5}$
(b) $x = 1$, $x = 4$; $y = -\sqrt{x}$
(c) $x = 0$, $x = 1$; $y = x^2 - 2x$
(d) $x = 0$, $x = 1$; $y = x^3 - x$
Answer: $\frac{1}{4}$
23. Find the areas of the figures lying between the x-axis and the graphs of
(a) $y = x^3 - x$
(b) $y = (x + 2)(x + 1)(x - 1)(x - 2)$.
24. Find the area of the figure bounded by the graphs of $y + 2x = 1$ and $x^2 = 1 - y$.
25. Find the areas of the figures bounded by the following pairs of curves.
(a) $x^2 = y + 1$ and $x^2 = 1 - y$
Answer: $\frac{8}{3}$
(b) $y = 4 - x^2$ and $x + y + 2 = 0$
Answer: 125/6
(c) $y = x - x^2$ and $y = x - 1$
(d) $y = x^2 - 4$ and $2x - y = 1$
(e) $y = x^3$ and $y = x$
26. Find the areas of the figures bounded by the following pairs of curves.
(a) $y^2 = 1 - x$ and $x = 0$
Answer: $\frac{4}{3}$
(b) $y^2 = x + 2$ and $y = x$
(c) $y^2 = 2$ and $2x + y = 2$
27. Prove the mean value theorem for integrals, namely, Theorem 6.3. Let f be a function continuous on $[a,b]$. Then there is a number $c \in [a,b]$ such that

$$\int_a^b f = (b - a)f(c).$$

Hints:
(a) The function f has an absolute maximum, $f(\bar{x})$, $\bar{x} \in [a,b]$ and an absolute minimum, $f(\underline{x})$, $\underline{x} \in [a,b]$, and therefore

$$(b - a)f(\underline{x}) \leq \int_a^b f \leq (b - a)f(\bar{x}).$$

(b) There is a number $c \in [a,b]$ at which

$$f(c) = \frac{1}{b-a} \int_a^b f.$$

(c) At which places in your proof did you use the continuity of f?

(d) Sketch a graph of some such function, and give a geometrical description of this theorem (think of areas).

28. Let f be a function differentiable on $[a,b]$, and π_n be an evenly spaced partition of $[a,b]$. Let x_j^* be any number in the jth subinterval of π_n, $x_{j-1} \leq x_j^* \leq x_j$. Suppose that Df is bounded:

$$|Df(x)| \leq N, \quad x \in [a,b].$$

Prove that

$$\left| \int_a^b f - \frac{1}{n}(b-a) \sum_{j=1}^n f(x_j^*) \right| \leq \frac{N}{n}(b-a)^2.$$

(If it is difficult to find a function G for the fundamental theorem, and if a computer is available, then the number

$$R_n = \frac{1}{n}(b-a) \sum_{j=1}^n f(x_j^*)$$

can be used for $\int_a^b f$ with a known maximum error.)

Hints:

(a) First show on a sample graph that for a nonnegative function, R_n is the sum of areas of vertical rectangles approximating the area under the graph.

(b) On each subinterval there is a c_j at which

$$\int_{x_{j-1}}^{x_j} f = (x_j - x_{j-1}) f(c_j) = \frac{1}{n}(b-a) f(c_j).$$

See Problem 27.

(c) Show that

$$\left| \int_{x_{j-1}}^{x_j} f - \frac{1}{n}(b-a) f(x_j^*) \right| = \frac{1}{n}(b-a) |f(c_j) - f(x_j^*)|$$

$$= \frac{1}{n}(b-a) |(c_j - x_j^*)\, DF(y_j)|, \quad y_j \text{ between } c_j \text{ and } x_j^*,$$

$$\leq \frac{N}{n^2}(b-a)^2,$$

and therefore

$$\left| \int_a^b f - \frac{1}{n}(b-a) \sum_{j=1}^n f(x_j^*) \right| \leq \frac{N}{n}(b-a)^2.$$

29. Apply Problem 28 to $\int_0^1 x^2\,dx$, as follows. Notice that $N = 2$.

(a) Take $n = 2$; find R_2 and the maximum error. You choose x_j^*.
(b) Take $n = 4$; find R_4 and the maximum error. You choose x_j^*.
(c) Compute $\int_0^1 x^2\,dx$ and compare with your approximation.

30. Repeat Problem 29 for each of

$$\int_0^4 x^2\,dx \quad \text{and} \quad \int_1^2 x^2\,dx.$$

31. For each of the integrals in Problems 29 and 30 find an n so that R_n differs from the integral by less than 1/100.

32. Devise a better approximating formula than R_n of Problem 28 as follows. Assume that D^2f is bounded.

$$|D^2f(x)| \leqslant M, \quad x \in [a,b]$$

(a) Define

$$T_n = \frac{1}{2n}(b-a)\sum_{j=1}^{n}[f(x_{j-1}) - f(x_j)], \text{ evenly spaced } \pi_n$$

so that

$$T_n = \frac{1}{2n}(b-a)[f(a) + 2\sum_{j=1}^{n-1} f(x_j) + f(b)]$$

and illustrate on the graph of a nonnegative function that T_n approximates the area under the graph by the sum of the areas of vertical trapezoids.

(b) By the mean value theorem

$$f(x_j) - f(x_{j-1}) = \frac{1}{n}(b-a)Df(s_j), \quad x_{j-1} < s_j < x_j,$$

and then the graph of the function ϕ,

$$\phi(y) = f(x_{j-1}) + (y - x_{j-1})Df(s_j), \quad x_{j-1} \leqslant y \leqslant x_j$$

is the top of one of the trapezoids you sketched in part (a). Also $\int_{x_{j-1}}^{x_j} \phi$ is the area of this trapezoid and

$$\sum_{j=1}^{n}\int_{x_{j-1}}^{x_j} \phi = T_n.$$

(c) Again by the mean value theorem

$$f(y) = f(x_{j-1}) + (y - x_{j-1})Df(t_j), \quad x_{j-1} < t_j < x_j, \quad x_{j-1} \leqslant y \leqslant x_j$$

and therefore

$$f(y) - \phi(y) = (y - x_{j-1})[Df(t_j) - Df(s_j)].$$

(d) Express this difference of values of the derivative of f by the mean value theorem,

$$Df(t_j) - Df(s_j) = (t_j - s_j)D^2f(v_j), \quad v_j \text{ between } t_j \text{ and } s_j,$$

so that

$$|(f-\phi)(y)| = |f(y) - \phi(y)| = |(y - x_{j-1})(t_j - s_j)D^2f(v_j)|$$
$$\leqslant (x_j - x_{j-1})^2 M = \frac{1}{n^2}(b-a)^2 M, \quad x_{j-1} \leqslant y \leqslant x_j.$$

(e) Therefore,

$$\left|\int_x^{x_j} f - \int_{x_{j-1}}^{x_j} \phi\right| = \left|\int_{x_{j-1}}^{x_j} (f-\phi)\right| \leqslant \frac{1}{n^3}(b-a)^3 M,$$

and

$$\left|\int_a^b f - T_n\right| \leqslant \frac{M}{n^2}(b-a)^3.$$

A sum T_n is called the *trapezoidal rule*.

33. Apply Problem 32 to $\int_0^1 x^2\,dx$ as follows. Notice that $M = 2$.
 (a) Take $n = 2$; find T_2 and the maximum error.
 (b) Take $n = 4$; find T_4 and the maximum error.
 (c) Compare these approximations with the correct value of the integrals.
34. Repeat Problem 33 for each of $\int_0^4 x^2\,dx$ and $\int_1^2 x^2\,dx$.
35. For each of the integrals in Problems 33 and 34 find an n so that T_n differs from the integral by less than 1/100. Compare with the corresponding values of n for R_n.
36. For the function $g(x) = 1/x$, $x > 0$, find a value of n such that the error using T_n for $\int_1^2 g$ is less than $\frac{1}{10}$. What is such an n using R_n? Compute T_n for this n. Try R_n if you are ambitious.
37. Let f be a function whose domain is the set of positive integers. Define $\lim_\infty f = L$ in case for any $\epsilon > 0$ there is an N such that $n > N$ implies $|f(n) - L| < \epsilon$. Show that the error in approximating an integral using T_n goes to 0 faster than the error using R_n, as n increases.

6.6 THE INTEGRAL AS A LIMIT OF INTERMEDIATE SUMS

For a partition π of the interval $[a,b]$, which is the domain of a function f, we form the sum

$$ {}_a^bS(\pi) = \sum_{j=1}^{n} f(x_j^*)(x_j - x_{j-1}), \quad x_{j-1} \leqslant x_j^* \leqslant x_j $$

over the subintervals of π. If each $f(x_j^*) = \max\{f(x)|x_{j-1} \leqslant x \leqslant x_j\}$, then ${}_a^bS(\pi) = {}_a^b\bar{S}(\pi)$, the upper sum. For any particular partition π,

$$ {}_a^b\underline{S}(\pi) \leqslant {}_a^bS(\pi) \leqslant {}_a^b\bar{S}(\pi), $$

so the intermediate sum is generally between the particular corresponding lower and upper sums. We now explain how the integral of a continuous function can be defined by means of intermediate sums, and we turn to some preliminary remarks in preparation.

DEFINITION 6.8 A function whose domain is an interval I is *uniformly continuous* (ufc) in case for any number $\epsilon > 0$ there is a number[3] $\delta > 0$ such that if $|x - y| < \delta$, with x and y in I, then $|f(x) - f(y)| < \epsilon$.

EXAMPLE The continuous function

$$ f(x) = \frac{1}{x}, \quad \frac{1}{2} < x \leqslant 1 $$

is ufc. Notice that for any $\epsilon > 0$, we may take $d = \epsilon/(2(2 + \epsilon))$, so that if $\frac{1}{2} < y < x < 1$, and $|x - y| < \delta$; then

$$ |f(x) - f(y)| = \left|\frac{1}{x} - \frac{1}{y}\right| = \frac{|x - y|}{xy} < 4|x - y| < \frac{\epsilon}{1 + \epsilon/2} < \epsilon. $$

EXAMPLE The continuous function

$$ f(x) = \frac{1}{x}, \quad 0 < x \leqslant 1 $$

is not ufc. To demonstrate why (see Definition 6.8) we must exhibit some $\epsilon > 0$ such that for any $\delta > 0$ whatever, there exist numbers x and y in dom f with $|x - y| < \delta$ and $|f(x) - f(y)| \nless \epsilon$.

We exhibit $\epsilon = 1$, and discuss the possibilities for $\delta < \frac{3}{2}$, leaving large numbers δ to you. Let $x = \delta/(2\delta + 3)$ and $y = 2\delta/3$. Clearly, $0 < x < y < 1$, so $x, y \in$ dom f, and

[3] This number δ is independent of the location of the numbers x and y in I as long as they differ by less than δ, which serves the interval I uniformly.

$$|x-y| = y - x = \frac{\delta}{3}\frac{4\delta+3}{2\delta+3} < \delta.$$

But then

$$|f(x)-f(y)| = \frac{1}{x} - \frac{1}{y} = 2 + \frac{3}{2\delta} > 1 = \epsilon.$$

Notice that for small δ we have only to choose x and y close to 0 to get a difference of more than 1 in the function values.

Thus a continuous function may or may not be ufc. However, if dom f is a closed interval we have the next lemma.

Lemma to Theorem 6.4: If f is a continuous function on a closed interval $[a,b]$, then f is ufc.

Proof: Choose $\epsilon > 0$. Define the set

$$C = \{c \mid c \in [a,b], \quad \text{restriction of } f \text{ to } [a,c] \text{ is ufc}\}.$$

Clearly, $a \in C$ and C is bounded, so the least upper bound axiom provides a number $c^* = \text{lub } C$. We now show that $c^* = b$. (See Figure 6.14.)

At c^*, f is continuous, and for some $\delta^* > 0$,

(A) $\quad x \in (c^* - \delta^*, c^* + \delta^*) \quad \text{implies} \quad |f(x) - f(c^*)| < \frac{\epsilon}{2}.$

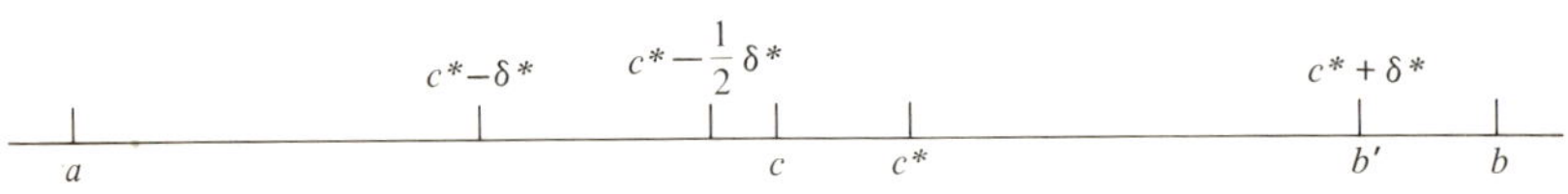

Figure 6.14

Also, since $c^* = \text{lub } C$, $c^* - \frac{1}{2}\delta^* \neq \text{ub } C$; that is, there is a number $c \in C$ for which

$$c^* - \frac{1}{2}\delta^* < c \leqslant c^*,$$

and

(B) $\quad f$ is ufc on $[a,c]$

Let δ_c be the number given by Definition 6.8, so that $|x-y| < \delta_c$ implies $|f(x)-f(y)| < \epsilon$.

Take $\delta = \min\{\frac{1}{2}\delta^*, \delta_c\}$, and let $b' = \min\{c^* + \delta^*, b\}$. For any two numbers x,y in $[a,b']$, for which $|x-y| < \delta$, there are two alternatives. First, either x or y, or both, are not in $[a,c]$. But then surely both are in $(c^* - \delta^*, c^* + \delta^*)$, and from line (A)

$$|f(x)-f(y)| \leqslant |f(x)-f(c^*)|+|f(y)-f(c^*)| < \frac{\epsilon}{2}+\frac{\epsilon}{2}=\epsilon.$$

Second, both x and y are in $[a,c]$ and from line (B),

$$|f(x)-f(y)| < \epsilon.$$

Therefore, f is ufc on $[a,b']$, and $b' \in C$. But $b' \geqslant c^* = \text{lub } C$, and so $b' = c^* \in C$. Of the two choices for b' we now see that $b' \neq c^* + \delta^*$, and so $b' = b \in C$. ■

Returning to the intermediate sums, recall that a partition creates a finite number of subintervals, so we can make the following definition.

DEFINITION 6.9 The *norm* of a partition π is

$$\|\pi\| = \max \{x_j - x_{j-1}| \text{ all subintervals of } \pi\}.$$

THEOREM 6.4 Let f be a continuous function on the closed interval $[a,b]$. Given any $\epsilon > 0$, there is a $\delta > 0$ so that if $\|\pi\| < \delta$, then

$$\left|\int_a^b f - {}_a^bS(\pi)\right| < \epsilon.$$

Proof: By Theorem 6.3, (Problem 27, Section 6.5) there is, for each subinterval of π, a number c_j so that

$$\int_{x_j}^{x_j} f=f(c_j)(x_j-x_{j-1}), \quad x_{j-1} \leqslant c_j \leqslant x_j.$$

and therefore,

$$\int_a^b f=\sum_{j=1}^n f(c_j)(x_j-x_{j-1}).$$

Choose $\epsilon > 0$. Then $\epsilon/(b-a) > 0$ and by the lemma, there is a number $\delta > 0$ so that if $x,y \in [a,b]$ and $|x-y| < \delta$, then $|f(x)-f(y)| < \epsilon/(b-a)$. Suppose that $\|\pi\| < \delta$. Then

$$\left|\int_a^b f - {}_a^bS(\pi)\right| = \left|\sum_{j=1}^n f(c_j)(x_j-x_{j-1}) - \sum_{j=1}^n f(x_j^*)(x_j-x_{j-1})\right|,$$

$$x_{j-1} \leqslant x^* \leqslant x_j,$$

$$= \left|\sum_{j=1}^n [f(c_j)-f(x_j^*)](x_j-x_{j-1})\right|$$

$$\leqslant \sum_{j=1}^n |f(c_j)-f(x_j^*)|(x_j-x_{j-1}),$$

$$|c_j - x_j^*| \leqslant \|\pi\| < \delta,$$

$$< \frac{\epsilon}{b-a}\sum_{j=1}^n (x_j-x_{j-1}) = \epsilon. \blacksquare$$

DEFINITION 6.10 Consider a collection of intermediate sums ${}_a^bS(\pi)$ for a function f with domain $[a,b]$. Suppose that for any $\epsilon > 0$ there is a $\delta > 0$ such that if $\|\pi\| < \delta$, then $|L - {}_a^bS(\pi)| < \epsilon$ for some number L. Then L is the *limit* of the intermediate sums as the norm goes to 0.

$$L = \lim_{\|\pi\| \to 0} {}_a^bS(\pi)$$

For a continuous function on a closed interval we have shown that this limit exists and is the integral

$$\int_a^b f = \lim_{\|\pi\| \to 0} {}_a^bS(\pi).$$

EXAMPLE Let f be the function

$$f: f(x) = \sqrt{1 - x^2}, \quad 0 \leqslant x \leqslant \frac{1}{\sqrt{2}}$$

whose graph (Figure 6.15) is an arc of the unit circle. The length of this arc can be approximated by the sum of the lengths of chords joining points on the arc consecutively. If π is a partition of the interval $[0, 1/\sqrt{2}]$, this approximation is

$$\sum_{j=1}^{n} \sqrt{(x_j - x_{j-1})^2 + (f(x_j) - f(x_{j-1}))^2}.$$

Since f is a continuously differentiable function, the mean value theorem yields

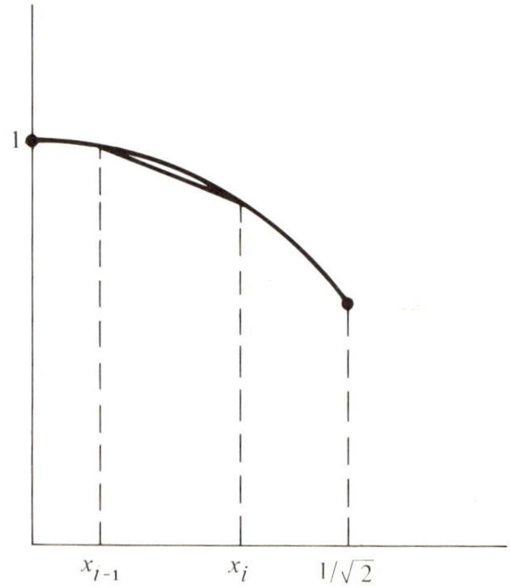

Figure 6.15

$$f(x_j) - f(x_{j-1}) = Df(x_j^*)(x_j - x_{j-1}), \quad x_{j-1} \leqslant x_j^* \leqslant x_j,$$

and so the approximation is, in fact, the intermediate sum

$$^{1/\sqrt{2}}_{0}S(\pi) = \sum_{j=1}^{n} \sqrt{1 + [Df(x_j^*)]^2}(x_i - x_{i-1}), \quad x_{j-1} \leqslant x_j^* \leqslant x_j.$$

If we take

$$\lim_{\|\pi\| \to 0} {}^{1/\sqrt{2}}_{0}S(\pi)$$

to be the length of the arc, then this length is[4]

$$L = \int_0^{1/\sqrt{2}} \sqrt{1 + (Df)^2} = \int_0^{1/\sqrt{2}} \frac{1}{\sqrt{1 - x^2}}\,dx.$$

We prove in Chapter 7 that $D \arcsin x = 1/\sqrt{1 - x^2}$, therefore

$$L = \arcsin \frac{1}{\sqrt{2}} - \arcsin 0 = \frac{\pi}{4}.$$

This approach to arc length applies to any continuously differentiable function f, and is contained in the next definition.

DEFINITION 6.11 If a function f has a continuous derivative Df on the closed interval $[a,b]$, then the *arc length* of (the graph of) f is

$$\int_a^b \sqrt{1 + (Df)^2}.$$

In applications to geometry it is usual to ask for the arc length of a geometric graph prior to the description of a function for which the graph is representative. Thus, in the example the question would be, "What is the length of one-eighth of the unit circle?"

EXAMPLE Find the circumference of the unit circle. Here, there is no single function whose graph is the whole circle, and we could resort to the previous example and multiply that result by 8. However, a parametric representation of the whole circle, as described in Chapter 4, page 121, can be used to solve the problem with one integration.

Describe the circle by the two functions, ϕ, ψ,

$$x = \phi(t) = \cos t, \quad y = \psi(t) = \sin t, \quad 0 \leqslant t \leqslant 2\pi.$$

The circumference of this circle can be approximated by the sum of the

[4] Since Df is continuous, $\sqrt{1 + (Df)^2}$ is ufc on the closed interval $[0,1/\sqrt{2}]$ and the integral exists.

lengths of chords joining points on the circle consecutively from $t=0$ to $t=2\pi$, as in Figure 6.16. If π is a partition of the closed interval $[0,2\pi]$, this approximation is

$$\sum_{j=1}^{n} \sqrt{[\phi(t_j)-\phi(t_{j-1})]^2+[\psi(t_j)-\psi(t_{j-1})]^2}.$$

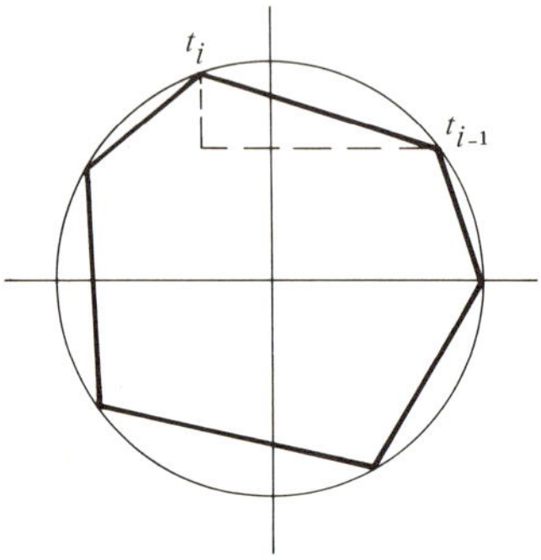

Figure 6.16

Since ϕ and ψ are continuously differentiable functions,[5] the mean value theorem for each of them gives

$$\begin{aligned}\phi(t_j)-\phi(t_{j-1}) &= D\phi(t_j^*)(t_j-t_{j-1}), \quad t_{j-1}\le t_j^*\le t_j\\ \psi(t_j)-\psi(t_{j-1}) &= D\psi(t_j^{**})(t_j-t_{j-1}), \quad t_{j-1}\le t_j^{**}\le t_j,\end{aligned}$$

where t_j^* and t_j^{**} are not necessarily equal. Hence the approximation of the circumference is given by

$$^{2\pi}_{0}T(\pi)=\sum_{j=1}^{n}\sqrt{[D\phi(t_j^*)]^2+[D\psi(t_j^{**})]^2}\,(t_j-t_{j-1}).$$

This is similar to the intermediate sum, except for the difference between t_j^* and t_j^{**}. In order to get an integral for the circumference of the circle we follow the pattern of the proof of Theorem 6.4.

Consider the integral[6] I

$$I=\int_0^{2\pi}\sqrt{(D\phi)^2+(D\psi)^2}.$$

By the mean value theorem for integrals (Section 6.5, Problem 27),

[5] This will be proved in Chapter 7.

[6] Since both ϕ and ψ are continuously differentiable on the closed interval $[0,2\pi]$, both $D\phi$ and $D\psi$ are ufc. In fact, so also is the function $\sqrt{(D\phi)^2+(D\psi)^2}$ ufc, as you should verify.

$$\int_{t_{j-1}}^{t_j} \sqrt{(D\phi)^2 + (D\psi)^2} = \sqrt{[D\phi(c_j)]^2 + [D\psi(c_j)]^2}(t_j - t_{j-1}), \quad t_{j-1} \leqslant c_j \leqslant t_j,$$

therefore

$$I = \sum_{j=1}^{n} \sqrt{[D\phi(c_j)]^2 + [D\psi(c_j)]^2}\,(t_j - t_{j-1}), \quad t_{j-1} \leqslant c_j \leqslant t_j,$$

see Figure 6.17.

By the triangle inequality,

$$\begin{aligned} &|\sqrt{[D\phi(c_j)]^2 + [D\psi(c_j)]^2} - \sqrt{[D\phi(t_j^*)]^2 + [D\psi(t_j^{**})]^2}| \\ &\quad \leqslant |\sqrt{[D\phi(c_j)]^2 + [D\psi(c_j)]^2} - \sqrt{[D\phi(t_j^*)]^2 + [D\psi(c_j)]^2}| \\ &\qquad + |\sqrt{[D\phi(t_j^*)]^2 + [D\psi(c_j)]^2} - \sqrt{[D\phi(t_j^*)]^2 + [D\psi(t_j^{**})]^2}|. \end{aligned}$$

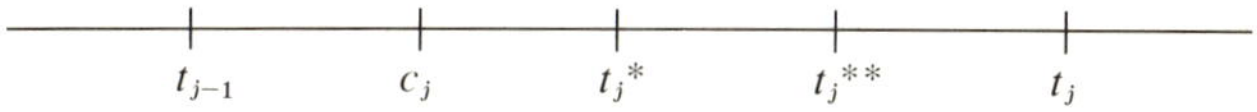

Figure 6.17

Choose any $\epsilon > 0$; then $(\epsilon/4\pi) > 0$, and by the uniform continuity of the function $\sqrt{(D\phi)^2 + (D\psi)^2}$, there is a $\delta > 0$ such that[7] if $\|\pi\| < \delta$,

$$\begin{aligned} \left|I - {}^{2\pi}_{0}T(\pi)\right| &= \Bigg|\sum_{j=1}^{n} [\sqrt{[D\phi(c_j)]^2 + [D\psi(c_j)]^2} \\ &\qquad - \sqrt{[D\phi(t_j^*)]^2 + [D\psi(t_j^{**})]^2}]\,(t_j - t_{j-1})\Bigg| \\ &\leqslant \sum_{j=1}^{n} |\sqrt{[D\phi(c_j)]^2 + [D\psi(c_j)]^2} \\ &\qquad - \sqrt{[D\phi(t_j^*)]^2 + [D\psi(c_j)]^2}|\,(t_j - t_{j-1}) \\ &\quad + \sum_{j=1}^{n} |\sqrt{[D\phi(t_j^*)]^2 + [D\psi(c_j)]^2} \\ &\qquad - \sqrt{[D\phi(t_j^*)]^2 + [D\psi(t_j^{**})]^2}|\,(t_j - t_{j-1}) \\ &< \sum_{j=1}^{n} \frac{\epsilon}{4\pi}\,(t_j - t_{j-1}) + \sum_{j=1}^{n} \frac{\epsilon}{4\pi}\,(t_j - t_{j-1}) \\ &= \frac{\epsilon}{4\pi}\,(2\pi + 2\pi) = \epsilon. \end{aligned}$$

Therefore,

$$I = \lim_{\|\pi\| \to 0} {}^{2\pi}_{0}T(\pi),$$

[7] Notice the two uses of the symbol π.

and if we take this limit to be the circumference of the unit circle, then the circumference is

$$I=\int_0^{2\pi}\sqrt{(D\phi)^2+(D\psi)^2}=\int_0^{2\pi}\sqrt{\sin^2 t+\cos^2 t}\;dt=\int_0^{2\pi}\underline{1}\;dt=2\pi.$$

The method we have just explained, which assigns a length to a curve represented parametrically, applies generally to *parametric* curves, and it is contained in the next definition.

DEFINITION 6.12 If a curve is described parametrically by the continuously differentiable functions, ϕ, ψ,

$$x=\phi(t),\quad y=\psi(t),\quad a\leqslant t\leqslant b,$$

then the *arc length* of the curve is

$$\int_a^b\sqrt{(D\phi)^2+(D\psi)^2}.$$

We know that the arc length integral of this definition may exist even though the functions ϕ and ψ do not have continuous derivatives $D\phi$ and $D\psi$. We could start with the following definition.

DEFINITION 6.13 A curve given parametrically by the functions ϕ and ψ is *rectifiable* (has length) in case

$$\int_a^b\sqrt{(D\phi)^2+(D\psi)^2}$$

exists.

Then we could construct a theory of rectifiable curves without depending on the continuous differentiability of the parametric functions.

6.6 PROBLEMS

1. Find the length of the parabolic segment $y=x^2$, $0\leqslant x\leqslant 1$. (See Problem 1(m), Section 8.5, for the integration.)
2. Find the length of the parabolic segment $y=\sqrt{x}$, $0\leqslant x\leqslant 1$.

$$\left(D[\sqrt{x(x+\tfrac{1}{4})}+\tfrac{1}{4}\log(\sqrt{x}+\sqrt{x+\tfrac{1}{4}})]=\sqrt{\frac{x+\tfrac{1}{4}}{x}}\right)$$

3. Show that Definition 6.11 recaptures the euclidean length of a line segment.
4. Find the length of the parabolic segment

$$x=t^2,\quad y=t,\quad -1\leqslant t\leqslant 2$$

$$(D[t\sqrt{t^2+\tfrac{1}{4}}+\tfrac{1}{4}\log(t+\sqrt{t^2+\tfrac{1}{4}})]=\sqrt{4t^2+1}).$$

5. Find the length of the segment of a semicubical parabola

$$x = t^2, \quad y = t^3, \quad 0 \leqslant t \leqslant 4.$$

6. Find the length of one arch of the cycloid,

$$x = t - \sin t, \quad y = 1 - \cos t, \quad 0 \leqslant t \leqslant 2\pi.$$

7. Write the arc length integral for the circumference of the ellipse

$$x = a \cos t, \quad y = b \sin t, \quad 0 < b < a, \quad 0 \leqslant t \leqslant 2\pi.$$

(A function whose derivative is this integrand is not shown at all.)

8. Prove that the arc length integral is invariant under a rotation. (See Appendix IV, page 278.)

REVIEW PROBLEMS

1. A certain function f is such that for any two numbers x and y,

$$|f(x+y) - f(x)| \leqslant y^2.$$

Show that there is a number K such that for all x,

$$f(x) = K.$$

2. If $x^2 = \int_a^x \phi$, what are possible functions ϕ, if any, and for each what are possible values of a?
3. An object moving in a straight line has an acceleration proportional to the square root of its velocity. It has an initial velocity of 16 feet per second and comes to rest in 6 seconds. How far does it go?
4. Two men are to share the work of pumping out the tank of Problem 1, Section 6.5. How much liquid should the first man pump out in order to do half the work?
Answer: $125\sqrt{2}\pi$ cubic feet
5. A slender bar of length 1 lies along the positive side of the x-axis, with the nearer extremity at a distance d from the origin O. If O is a center of attraction such that the attraction on every particle of the bar varies inversely as the square of its distance from O, find the total attraction.
6. Find the area between the graph of $y = x^3 - 12x$ and the tangent drawn to it at the maximum point on the curve.
Answer: 108
7. Find a if $y = a$ divides into two equal parts the area bounded by the parabola $y = 4x - x^2$, the x-axis, and $x = 2$.
8. Find the area bounded by the curve

$$\sqrt{x} + \sqrt{y} = 1$$

and the coordinate axes.

9. Show that the curve $y = x^n (n > 0)$ divides the unit square bounded by $x = 0$, $x = 1$, $y = 0$, and $y = 1$ in the ratio $n:1$.
10. An area in the first quadrant is bounded by the curve $x^2y = 6x^3 + 8$, the x-axis, and the lines $x = k$ and $x = k + 1$. Find the minimum value of the area.
    Answer: $k = 1$
11. A segment of a parabola is the area bounded by the parabola and a chord perpendicular to the axis. Prove that the area of any segment is equal to two-thirds the area of the circumscribed rectangle.
12. Consider all the parabolas which have their axes parallel to the y-axis, open down and pass through (0,0) and through the point (1,2). Find an equation for that one which, together with the x-axis, bounds a figure of least area.
13. Show that if ϕ and ψ are functions, both of which are continuous on the interval $[a,b]$, and if $\phi(x) \leqslant \psi(x)$ for each x in $[a,b]$, then

$$\int_a^b \phi \leqslant \int_a^b \psi.$$

14. Let f and g be functions which are continuous on an interval $[a,b]$, and let g be such that $g(x) \geqslant 0$ for each x in $[a,b]$. Show that there is a number c in $[a,b]$ such that

$$\int_a^b fg = f(c) \int_a^b g.$$

    Hint: Let m and M be the minimum and maximum values of f on $[a,b]$. Then $mg(x) \leqslant f(x)g(x) \leqslant Mg(x)$. Now use Problem 13 above.

Chapter 7 ELEMENTARY FUNCTIONS

7.1 AN INTEGRATION PROBLEM

The problem of integration is now before us. That is, how do we express $\int_a^b f$ by means of standard formulas with which we are familiar. These formulas should describe functions whose properties are not too hard to determine. Otherwise we must be satisfied with a knowledge of the existence of $\int_a^b f$, based on the theoretical work of the last chapter, and approximating processes such as the trapezoidal rule. In fact, these approximating processes are now very usable for applications, since modern computers are usually available, but it will deepen our understanding of the integral to attempt its computation without a machine. In Chapter 8 the technique of integration will be discussed, and the functions constructed in this chapter will be used.

Consider $\int_1^x t^n \, dt$, where n is an integer, for instance. If $n \neq -1$, we have seen that if $g(t) = \frac{1}{n+1} t^{n+1}$, then $Dg(t) = t^n$, and by the fundamental theorem,

$$\int_1^x t^n \, dt = \frac{1}{n+1} (x^{n+1} - 1), \quad x > 0.$$

If $n = -1$, $\int_1^x (1/t) \, dt$ requires special treatment. Suppose that there were a rational function R such that

$$\text{(A)} \qquad \int_1^x \frac{1}{t} \, dt = R(x), \quad x > 0,$$

where $R(x) = \frac{P(x)}{Q(x)}$. Denote the degrees of the polynomials by $\deg P = p$, $\deg Q = q$. Differentiate line (A) to find

$$Q^2(x) = xQ(x) \, DP(x) - xP(x) \, DQ(x) = S(x),$$

say, where if $p \neq q$,

$$\deg Q^2 = 2q = \deg S = p + q,$$

a contradiction. On the other hand, if $p = q$, then

$$\deg Q^2 = 2q = \deg S < p + q = 2q,$$

a contradiction again.

Therefore, there is no rational function R which will serve to represent $R(x) = \int_1^x (1/t) \, dt$. Perhaps some other familiar function will, but we find it easier to abandon the search and to examine the integral itself. Since the function r,

$$r(t) = \frac{1}{t}, \quad 0 < t$$

is continuous, we know that the function s,

$$s(x) = \int_1^x r, \quad 0 < x,$$

exists, and we proceed to study it. We give it the name *log*,[1] and henceforth write

$$\log x = \int_1^x r = \int_1^x \frac{1}{t} \, dt, \quad 0 < x.$$

[1] This is a contraction of logarithm.

7.2 THE LOGARITHM FUNCTION AND THE EXPONENTIAL FUNCTION

THEOREM 7.1 If $x > 0$, and $y > 0$, then

$$\log(xy) = \log x + \log y.$$

Proof: Define, for each $x > 0$, the function ${}_xg$,

$${}_xg(y) = xy, \quad y > 0$$

and the function ${}_xf$,

$${}_xf(y) = \log xy = \log \circ {}_xg(y).$$

Differentiate ${}_xf$ by the chain rule,

$$D_xf = [(D \log) \circ {}_xg]D_xg,$$

and evaluate this formula at the number y. Piecemeal, we have

$$D_xg(y) = x, \quad (D \log) \circ {}_xg(y) = \frac{1}{{}_xg(y)} = \frac{1}{xy}.$$

Therefore

$$D_xf(y) = \frac{1}{y} = D \log y,$$

and so

$${}_xf(y) = \log y + {}_xk(y), \quad x > 0,$$

where ${}_xk$ is a constant function. At $y = 1$, $\log y = \log 1 = 0$, since $\log 1 = \int_1^1 (1/t)\, dt = 0$, and

$${}_xk(y) = {}_xk(1) = {}_xf(1) = \log x,$$

so that

$$\log xy = {}_xf(y) = \log y + \log x. \blacksquare$$

Corollary to Theorem 7.1: For $x > 0$ and any integer n,

$$\log x^n = n \log x.$$

Proof: If $n = 0$, then $\log x^n = \log 1 = 0 = n \log x$. Also, $\log x^{n+1} = \log x\, x^n = \log x + \log x^n$ by Theorem 7.1, and if $\log x^n = n \log x$, then

$$\log x^{n+1} = (n + 1) \log x.$$

This induction on n proves the formula for $n \geqslant 0$.

If $n < 0$, then $-n > 0$. Initially, note that

$$0 = \log 1 = \log(x\, x^{-1}) = \log x + \log x^{-1}.$$

Hence, $\log x^{-1} = -\log x$. Generally, then,

$$\begin{aligned}\log x^n &= \log(x^{-n})^{-1} = -\log x^{-n}, \quad -n > 0\\ &= -(-n)\log x, \quad \text{by the above induction}\\ &= n \log x. \blacksquare\end{aligned}$$

Lemma to Theorem 7.2: The logarithm function is strictly monotone increasing, and range log is the set of real numbers.

Proof: Since the function r,

$$r(t) = \frac{1}{t}, \quad t > 0$$

is strictly monotone decreasing, the minimum value of r on a closed interval $[1,x]$ is $1/x$, and the maximum value is 1. By Lemma 3 to Theorem 6.1,

$$\frac{1}{x}(x-1) \leqslant \log x = \int_1^x \frac{1}{t}\,dt \leqslant x - 1, \quad x \geqslant 1.$$

Thus, we see that $\log x > 0$ for $x > 1$.

In particular, $\log 2 > 0$. Choose any number A and a positive integer $n > A/(\log 2)$. Then $\log 2^n = n \log 2 > A$, so range log is unbounded above. For any number A, again, take n as before and $\log 2^{-n} = -n \log 2 < -A$, so range log is unbounded below.

Consider any $y > 0$. In the last paragraph we observed that there are numbers $\bar{y}$ and $\underline{y}$ such that

$$\log \underline{x} = \underline{y} < y, \quad \log \bar{x} = \bar{y} > y$$

for some $\underline{x}$ and $\bar{x}$ (powers of 2). Since log is continuous, the intermediate value lemma says that there is a number x at which $\log x = y$. Thus, range log is the set of real numbers.

Suppose that $0 < x < y$; then the mean value theorem says that there is a number c, $x < c < y$, at which

$$\log y - \log x = (D \log c)(y - x) = \frac{1}{c}(y - x) > 0,$$

and so $\log x < \log y$. Thus, log is strictly monotone increasing. ■

THEOREM 7.2 The logarithm function has an inverse, which we call the *exponential* function, abbreviated exp.

Proof: The lemma did the work because of Theorem 3.17. ■

It is now easy to visualize the exponential function. Very quickly,

dom exp = range log = set of real numbers;
range exp = dom log = set of nonnegative real numbers;
$D \exp = D \log^{-1} = \dfrac{1}{(D \log) \circ \log^{-1}} = \log^{-1} = \exp;$
$\exp 0 = \exp \circ \log 1 = I(1) = 1;$
exp is strictly monotone increasing and differentiable.

There is a linguistic reason for the name *exponential,* related to the word *exponent.* Let us rephrase Theorem 7.1.

THEOREM 7.3 For any two numbers x, y,

$$\exp(x + y) = \exp x \exp y.$$

Proof: Let $\exp x = X$, $\exp y = Y$; then $x = \log X$, $y = \log Y$, and

$$\begin{aligned}\exp(x + y) &= \exp[\log X + \log Y] \\ &= \exp(\log XY) \\ &= XY = \exp x \exp y. \blacksquare\end{aligned}$$

Since we know dom exp contains all numbers, in particular it contains 1; that is $\exp 1 = e$ for some number e. Now

$$\exp 0 = 1 = e^0$$

and if $\exp n = e^n$, $n > 0$ an integer, then

$$\exp(n + 1) = \exp n \exp 1 = e^n e = e^{n+1}.$$

So by this induction on n, clearly, the values of exp at positive integers are integral powers of e. For negative integers $n < 0$, note that

$$1 = \exp 0 = \exp(n + (-n)) = \exp n \exp(-n)$$

and hence, again

$$\exp n = \frac{1}{\exp(-n)} = \frac{1}{e^{-n}} = e^n.$$

Finally, if p is an integer, then there is a value of $\exp(1/p) = N$, say. Therefore, $(1/p) = \log N$, and $\log N^p = 1$, $N^p = e$, so that

$$\exp \frac{1}{p} = N = e^{1/p}.$$

The exponential function recaptures the ordinary algebraic power function through rational exponents. But it can do more. There is no purely algebraic way to define e^k, where k is irrational. However, $\exp k$ exists for any real number k, and we make the next definition.

DEFINITION 7.1 For any real number k, the kth *power* of e is $\exp k$; that is,

$$e^k = \exp k.$$

DEFINITION 7.2 For any $b > 0$, and any real number k, the kth power of b is $\exp(k \log b)$; that is,

$$b^k = \exp(k \log b).$$

This can be called the exponential *to the base b,*

$$\exp_b\colon \exp_b x = b^x, \quad b > 0, \quad x \text{ real},$$

but this terminology is not customarily used.

From Definition 7.2 we get the familiar logarithmic formula

$$\log b^k = k \log b, \quad b > 0, \quad k \text{ real}.$$

As a final comment about the general logarithmic and exponential functions, we introduce the next definition.

DEFINITION 7.3 The logarithm *to the base b*,[2] $b > 0$, $b \neq 1$, of a number x is $\log_b x$, where

$$\log_b x = \frac{\log x}{\log b}, \quad x > 0.$$

The definition is made in this rather curious way so that the exponential function to the base b and the logarithm function to the base b are inverses. To see this, write

$$\begin{aligned}(\log_b \circ \exp_b)(x) &= \log_b(\exp_b x)\\ &= \log_b(\exp(x \log b))\\ &= \frac{1}{\log b}\log(\exp(x \log b))\\ &= \frac{1}{\log b}(x \log b) = x, \quad x \text{ real}.\end{aligned}$$

7.2 PROBLEMS

1. Sketch a graph of log after verifying the following facts.
   (a) $\log 1 = 0$
   (b) $\log x < 0$ for $x < 1$, $\log x > 0$ for $x > 1$
   (c) $D \log x > 0$, $x > 0$
   (d) $D^2 \log x < 0$, $x > 0$
   (e) For any number $\epsilon > 0$ there is an X such that if $x > X$, then $D \log x < \epsilon$. $(\lim_{x\to\infty} D \log x = 0)$
   (f) For any number A there is an X such that if $0 < x < X$, then $D \log x > A$. $(\lim_{x\to 0} D \log x = \infty)$
   (g) $D \log 1 = 1$
2. Use the results of Problem 1 to sketch a graph of exp.
3. Verify the properties of exp listed under Theorem 7.2.
4. Using the Definition 7.2, prove the *laws of exponents*,
   (a) $b^x b^y = b^{x+y}$
   (b) $(b^x)^y = b^{xy}$, x, y real numbers.

[2] In some books log means $\log_{10}$, and ln means $\log_e = \log$.

5. Prove the following.
   (a) $\log_b(xy) = \log_b x + \log_b y$
   (b) $\log_b x^k = k \log_b x$
6. Show that the differential of exp at 0, $d\,\exp(0,h)$, is h. Prove that $e > 2$. (For problems take $e \cong 2.7$.)
7. We know that $e^k = \exp k$ by definition; that is, that $\exp_e = \exp$. Show that $\log_e = \log$.
8. Prove that $\log_a b \log_b a = 1$.
9. Prove that $\log_a x = \log_a b \log_b x$.
10. Verify the formulas; state the numbers for which they are valid.
    (a) $D \log_b x = (1/x) \log_b e$ (b) $D \log x = (1/x)$
    (c) $Db^x = b^x \log b$ (d) $De^x = e^x$
11. Find the derivative of each of the following functions.
    Hint: Make liberal use of the chain rule.
    (a) $\log(1 - x^3)$
    (b) $\log(1 - x)^3$ (Remember that $\log y^n = n \log y$.)
    (c) $\log \sqrt{4 - x^2}$
    (d) $\log(x + 1)(x^2 - 2)$ (Remember Theorem 7.1.)
    (e) $\log \sqrt{(x + 1)/(x - 1)}$
    (f) $\log \sqrt{x^5 + 5}$
    (g) $\log x \sqrt{1 \quad x}$
    (h) $\log(x^2 \log x)$
    (i) $\log 1/\sqrt{x}$
    (j) $\log[(x^2 - 1)/\sqrt{x^2 + 1}]$
    (k) $\log \left[\dfrac{\log x}{\sqrt{x}}\right]$
12. Find the derivative of each of the following functions.
    (a) $\dfrac{\sqrt{x + 1}}{x^2 - 1}$

    Hint: Set $f(x) = \dfrac{\sqrt{x + 1}}{x^2 - 1}$; then $\log f(x) = \frac{1}{2}\log(x + 1) - \log(x^2 - 2)$,

    and $\dfrac{1}{f(x)} Df(x) = \dfrac{1}{2(x + 1)} - \dfrac{2x}{x^2 - 1}$. Solve for $Df(x)$.

    (b) $\sqrt{\dfrac{x + 1}{x + 1}}$
    (c) $x(x^2 + 2)(x^4 + 4)$
    (d) $\sqrt{x}\sqrt[3]{1 - x}$
    (e) $\dfrac{3x^2 + 3}{x\sqrt{x^2 - 4}}$
    (f) $\left[\dfrac{(x - 1)(x + 2)}{(x - 2)(x + 3)}\right]^{1/2}$
    (g) $\dfrac{\sqrt[3]{x^2 + 1}\sqrt[4]{x^3 + 2}}{\sqrt[5]{x^4 + 3}}$
    (h) $(1 - x^2)^{5/2}(1 - x^3)^{5/3}(1 - x^6)^{5/6}$

13. Find the derivatives of each of the following functions.
 (a) $\log_{10} x$ (b) $\log_{10}(1 - x^2)$
 (c) $\log_2(\sqrt{x}\,\sqrt[3]{(1-x)})$ (d) $\log_3(1 + x)/(1 - x)$
 (e) $\log_{101}[1 + \log_{102}(1 + x)]$
14. Find the angle of intersection of the graphs of $\log x$ and $\log 2x$.
15. Find the equation of the tangent line to the graph of $\log x$ that
 (a) has slope 1
 (b) contains the origin.
16. Find the extreme values of each of the following.
 (a) $x \log x$ (b) $\dfrac{x}{\log x}$
 (c) $\dfrac{1 - \log x}{x}$ (d) $\log(1 + x^2)$
 (e) $\log(1 - x^2)$
17. Prove that if $f(x) = x^r$, $x > 0$, r real, then $Df(x) = rx^{r-1}$.
18. Let $f(x) = x^r$, $x > 0$, r real, and $g(x) = r^x$, $r > 0$, x real. Find the following.
 (a) Df (b) Dg
 (c) $D(fg)$ (d) $D(f + g)$
 (e) $D(f \circ g)$ (f) $D(g \circ f)$
19. Find the derivative of each of the following functions.
 (a) e^{4x} (b) e^{4x^2}
 (c) e^{-x^2} (d) $e^{x/(1+x)}$
 (e) $e^{\sqrt{x}}$ (f) $\sqrt{e^x}$
 (g) $(e^{\sqrt[3]{y}} + 1)^2$ (h) $\dfrac{e^s - e^{-s}}{e^s + e^{-s}}$
 (i) $\log(e^{x/2} + 2x)$ (j) $e^{(\log w + 2w)}$
 (k) $\dfrac{(e^t - 1)\sqrt{1 + e^{t^2}}}{1 - 2e^{-t}}$ (l) $\log we^w$
 (m) $\log\left(\dfrac{1}{w}\right) e^{-w}$ (n) $\dfrac{e^{s^2+s}}{s^2 + s}$
 (o) $e^{1/y}\log(y^e)$ (p) $x^2e^{x^2}$
 (q) $ye^{1/y}\sqrt{e^{2y} + \log y}$ (r) $\dfrac{e^{-2x}}{(1 - e^x)^2}$
 (s) $(1 + w)(1 + e^w)(1 + \log w)$
20. Find the derivative of each of the following.
 (a) $b^2(e^{x/b} - e^{-x/b})$ (b) $e^{x^3} + 1$
 (c) $e^{\sqrt{1-x}}$ (d) $\sqrt{x}\,e^{\sqrt{x}}$
 (e) $\dfrac{x}{e^{x^2}}$
21. Find the derivative of each of the following.
 (a) 2^x (b) 2^{x^2}
 (c) $\sqrt{1 - 10^{2x}}$ (d) $\log(1 + 3^x)$

(e) $\dfrac{2^{x^2} + 3^{-x^3}}{4^{x^4}}$ (f) $5^{\log x}$

(g) $6^x + x^6$

22. Find the maximum, minimum, and inflection points of each of the following functions, and sketch their graphs.
(a) $f(x) = e^{x^2}$ (b) $f(x) = e^{-x^2}$
(c) $g(x) = e^{-x}$ (d) $h(t) = te^t$
(e) $\phi(w) = e^{-w^3}$ (f) $h(y) = e^{1/y}$
(g) $\phi(x) = \frac{1}{2}(e^x + e^{-x})$ (h) $g(x) = \frac{1}{2}(e^x - e^{-x})$
(i) $h(x) = \dfrac{e^x + e^{-x}}{e^x - e^{-x}}$

23. Find the area of the largest triangle cut from the first quadrant by a line tangent to the curve $y = e^{-x}$.
Answer: $2/e$.

24. Find the largest rectangle which has its base on the x-axis, and its other two vertices on the graph of $y = e^{-x^2}$. Show that the two upper vertices are at the inflection points of the curve.

25. Find Df if $y = f(x)$ where
(a) $e^{xy} = e^x + e^y$ (b) $xy = e^{x/y}$
(c) $\log_{10}(x + y) = \log_2(e^x + e^{-y})$
(d) $ye^x - xe^y = \dfrac{xy}{\log(xy)}$.

26. Find the equations of the line through the origin which is tangent to the graph of each of the following.
(a) $y = e^x$ (b) $y = e^{2x}$
(c) $y = e^{-x}$ (d) $y = e^{x^2}$

27. Where do the curves $y = e^{-x}$ and $y = e^{-x^2}$ intersect? What is the tangent of the angle between the two tangent lines at this point?

28. Find Df if $f(x) = x^x$.

29. State the domain of and find Df if $f(x) =$
(a) $(1 - x)^x$ (b) x^{1-x}
(c) $(\log_e x)^x$ (d) $x^{\log x}$
(e) $x^{1/x}$ (f) $(10x)^{3x}$
(g) $(x^2 - 1)^e$ (h) $(\log x)^{\log x}$.

30. Find the minimum value of

$$f(x) = x^x, \quad x > 0.$$

31. Show that exp "goes to infinity" faster than I, the identity function whose domain is the positive real numbers. By this we mean that for any number ϵ there is a number N such that if $y > N$, then $(y/\exp y) < \epsilon$.
Hints:
(a) For each pair of numbers x, y, with $0 < x < y$, there is a

number c, $x < c < y$ such that

$$\frac{1}{\exp c} = \frac{DI(c)}{D\exp c} = \frac{I(y) - I(x)}{\exp y - \exp x} = \frac{y - x}{\exp y - \exp x},$$

by the double mean value theorem (review Problem 22, Chapter 5).

(b) Therefore,

$$\frac{y}{\exp y} = \frac{1}{\exp c}\,\frac{1 - \exp(x - y)}{1 - \dfrac{x}{y}}, \quad 0 < x < c < y$$

$$= \frac{1}{\exp c}\,\frac{1 - \exp x(1 - x)}{1 - \dfrac{1}{x}}, \quad \text{in particular for } y = x^2.$$

(c) Now think of y as very "large."

32. Show that exp goes to infinity faster than I^n for any positive integer n; that is, exp "grows" faster than any power function.

Hints:

(a) Following the pattern of Problem 31 (a), prove that for any two continuously differentiable functions f and g on the real numbers, if (make an appropriate definition of $\lim_\infty f = \infty$)

(i) $\lim_\infty f = \infty, \quad \lim_\infty g = \infty$

(ii) $Dg(y) \neq 0, \quad y$ real

(iii) $\lim_\infty (Df/Dg) = 0$

then $\lim_\infty (f/g) = 0$.

(b) With the result of (a), show that if $\lim_\infty \dfrac{I^n}{\exp} = 0$, then $\lim_\infty \dfrac{I^{n+1}}{\exp} = 0$.

(c) Use an induction on n, with Problem 31, to show that, in fact,

$$\lim_\infty \frac{I^n}{\exp} = 0, \quad n = 1, 2, 3, \ldots .$$

33. Show that any root function ${}_nR$

$${}_nR(x) = \sqrt[n]{x}, \quad x > 0, \quad n \text{ a positive integer},$$

goes to infinity faster than log.

Hint: Examine $\lim_\infty (\log/{}_nR)$ by the substitution $\log x = y$, so that

$$\left(\frac{\log x}{\sqrt[n]{x}}\right)^n = \frac{y^n}{\exp y},$$

and apply Problem 32.

34. Prove that for each triple of numbers k, a, A, there is, in the class of nonvanishing differentiable functions, just one function f for which

$$Df = kf, \quad f(a) = A.$$

Hints:

(a) Either $f(x) > 0$, or $f(x) < 0$ for all $x \in \operatorname{dom} f$, an interval.

(b) If $f(x) > 0$, $x \in \operatorname{dom} f$, then $D \log f = Df/f$, and if $f(x) < 0$, $x \in \operatorname{dom} f$, then $D \log(-f) = (Df/f)$.

(c) $$\log|f(x)| - \log|f(a)| = \int_a^x (Df/f) = \int_a^x k\,dx$$
$$= k(x-a),\ x \in \operatorname{dom} f.$$

(d) $f(x) = A \exp k(x-a)$, $x \in \operatorname{dom} f$.

(e) In this problem you solved the *differential equation* $Df = kf$, with the *initial condition* $f(a) = A$.

35. A radioactive substance decomposes at a rate proportional to the amount present. Suppose that 10 percent of the original amount has decomposed at the end of 2 years. When will one-half of the original amount have decomposed?

Hints:

(a) Let a nonvanishing differentiable function f, with domain $t > 0$, measure the amount of substance present at time t, so that $f(0) = A$.

(b) Interpret the physical description to mean $Df = kf$ for some k.

(c) Apply Problem 34 to show that

$$f(t) = A \exp kt, \quad t > 0,$$

where $f(2) = A \exp 2k = 0.9A$, so that $k = \frac{1}{2} \log 0.9$.

(d) When $f(t) = (\frac{1}{2})A$, then,

$$\frac{1}{2} = \exp\left(\frac{1}{2}\, t \log 0.9\right)$$

so that $t = (2 \log \frac{1}{2})/(\log 0.9) \cong 11.8$.

(e) From observing the substance for 2 years, you have computed its half-life, assuming it decomposes according to a nonvanishing differentiable function.

36. Find the half-life of radium if 4 percent decomposes in 100 years.
Answer: about 2000 years

37. When a capacitor discharges to ground through a resistance, the charge on the capacitor decreases at a rate proportional to the charge. If the original charge is 100 coulombs and if the charge at the end of 3 seconds is 70 coulombs, what was the charge at the end of 1 second?

38. If bacteria increase in number at a rate proportional to the number present, how long will it take before 1,000,000 bacteria increase to 10,000,000, if it takes 12 minutes to increase to 2,000,000?

39. The population of a city increases at a rate proportional to the population. In 20 years the population increases from 15,000 to 30,000. How long will it take to double in size again?

40. Light is absorbed by a certain kind of glass in such a way that the

intensity of the transmitted light decreases at a rate proportional to the thickness of the glass. If the intensity is diminished by 50 percent by 10 inches of glass, how much glass is needed to prevent all but 1 percent of the light from passing through?

41. The rate at which the temperature of a heated body decreases is proportional to the difference between its temperature and that of the air around it. At an air temperature of 60° how long will it take a body to cool from 120° to 80° if it takes 1 hour to cool from 150° to 90°?
42. For which values of the number a will the graph of $y = a - e^x$ intersect the x-axis and for which will it not? For which values of the number b will the graph of $y = bx^2 - 5x - e^x$ have a point of inflection and for which will it not?
43. (a) Draw the graph of $y^{\log_e x} = 1$.
    (b) Express $a^{\log_e x}$ as a power of x.
44. For what value of a is the curve $y = a^x$ tangent to the line $y = x$?
45. Prove that the curve $y = e^x + x^3$ intersects the x-axis in exactly one point.
46. Find and simplify the derivative of f if

$$f(x) = \left[\log(x+1) + \sqrt{2x+x^2} - \frac{4}{x+\sqrt{2x+x^2}}\right].$$

47. At what point on the curve $y = e^{-x^2}$ is the slope greatest?
48. Find the point of inflection of the curve $xy = \log_e x$.
49. Show that the curve $y = x \log_e x$ is everywhere concave upward. Find the minimum point and sketch the curve.
50. Sketch the graphs of the following.
    (a) $y = 1/(e^x - 1)$
    (b) $y = 1/(e^{1-x})$
51. The line AB is tangent to the curve $y = e^x + 1$ at the point B, and crosses the x-axis at A. Find the coordinates of the point of tangency B which make the length of the segment AB a minimum.
52. Find the volume generated by revolving about the x-axis the area bounded by the x-axis, the curve $y = (\log_e x)/\sqrt{x}$, and the ordinate through the point where y has an extreme value.
    Answer: $8\pi/3$
53. What is the length of the curve

$$y = \sqrt{a^2 - x^2} - a \log_e \left[\frac{a + \sqrt{a^2 - x^2}}{x}\right]$$

from $x = a/4$ to $x = a$?

54. The rate at which a liquid is flowing out of a vessel at any instant is proportional to the amount left in it at that instant. If the vessel is half empty in 1 hour, when will it be three-fourths empty?

55. A vertical cylindrical tank full of water has a leak at the bottom. If the water runs out at a rate proportional to its depth and if 10 percent of the water escapes the first day, how long will be required for 75 percent to escape?
56. The rate at which water runs out of a tank through a horizontal outlet is proportional to the square root of the height of the surface of the water above the outlet. If the tank is cylindrical and if half the water runs out in 30 minutes, when will it be empty?

57. DEFINITION $\lim_{n \to \infty} f(n) = A$ if for any $\epsilon > 0$ there is a number N such that when

$$n > N,$$

then

$$|f(n) - A| < \epsilon.$$

Prove that

$$\lim_{n \to \infty} \log \left(1 + \frac{1}{n}\right)^n = 1.$$

Hint: $\log \left(\left(1 + \frac{1}{n}\right)^n\right) = n \left(\log \left(1 + \frac{1}{n}\right)\right) = n \int_1^{1+(1/n)} \frac{dx}{x}.$ Now use Lemma 3 to Theorem 6.1.

58. Assuming that $\lim_{n \to \infty} (1 + (1/n))^n = e$ (the number such that $\log e = 1$), use the definition of the derivative to show that

$$D \log x = \frac{1}{x}.$$

59. (a) Show that

$$\lim_{n \to \infty} \left(1 + \frac{1}{n}\right)^n = e.$$

(b) Calculate the value of $(1 + (1/n))^n$ for $n = 1,2,3,4,5$.

7.3 ANOTHER INTEGRATION PROBLEM

Continuing a pursuit of a technique of integration, consider the integral $\int_0^x \frac{1}{1+t^2}\,dt$. This is a natural extension of the integration problem posed by $\int_1^x \frac{1}{t}\,dt$ to a simple rational function with a quadratic denominator. Possibly we can do it with rational functions.

Suppose that for some polynomial functions P, Q we set $R = \frac{P}{Q}$ and try to solve

$$DR(x) = \frac{1}{1+x^2}, \quad x \text{ real},$$

in which, we assume, P and Q have no common irreducible polynomial factor. Differentiating R, we find that

$$\text{(A)} \qquad Q^2(x) = (1+x^2)[Q(x)DP(x) - P(x)DQ(x)]$$

and since $1 + x^2$ is irreducible (does not factor into linear polynomials with real coefficients), the polynomial Q has a factor $1 + x^2$,

$$Q(x) = (1+x^2)Q_1(x), \quad \deg Q_1 = \deg Q - 2.$$

Therefore,

$$DQ(x) = 2xQ_1(x) + (1+x^2)DQ_1(x),$$

and (A) implies

$$2xQ_1(x)P(x) = (1+x^2)[(DP(x)+1)Q_1(x) - DQ_1(x)],$$

so that the polynomial Q_1 has a factor $1 + x^2$,

$$Q(x) = (1+x^2)Q_1(x) = (1+x^2)^2Q_2(x), \quad \deg Q_2 = \deg Q - 4.$$

Repeating this process a finite number of times, we find that if $q = 2m$, even, then

$$Q(x) = a(1+x^2)^m$$

so that line (A) implies that

$$2mxP(x) = (1+x^2)[a(1+x^2)^{m-1} + DP(x)],$$

a contradiction, since P has no factor $1 + x^2$. If q is odd, $q = 2m + 1$,

$$Q(x) = (ax+b)(1+x^2)^m,$$

and line (A) gives

$$2mx(ax+b)P(x) = (1+x^2)[(ax+b)(DP(x) - (1+x^2)^{m-1}(ax+b)) - aP(x)],$$

with the same contradiction.

Therefore, there is no rational function whose derivative is $1/(1+x^2)$. Had we constructed the calculus with the complex number system, we would see that the function f,

$$f(x) = \frac{1}{2i}\log\frac{1+ix}{1-ix}, \quad i = \sqrt{-1}, \quad x \text{ real},$$

takes a real value for each real x, and that

$$DF(x) = \frac{1}{1+x^2}.$$

(You can check this if you assume that the formal rules for differentiation hold when complex coefficients are present, but restricted to the real number system, as we are, this formula is denied us.) Rather than continue the search in the class of familiar functions for the function $\int_0^x \frac{1}{1+t^2}\,dt$, we give it a name, *arctan,* and study its properties directly.

$$\arctan x = \int_0^x \frac{1}{1+t^2}\,dt, \quad x \text{ real}.$$

Since the function described by $1/(1+x^2)$ is continuous with the real numbers for its domain, the integral exists for all x, and this justifies the domain indicated.

7.4 THE CIRCULAR (TRIGONOMETRIC) FUNCTIONS

If 0 is interior to the domain (an interval) of a function f, and if $-x \in \operatorname{dom} f$ whenever $x \in \operatorname{dom} f$, then we say

$$f \text{ is } \textit{odd} \text{ in case } f(-x) = -f(x)$$

and

$$f \text{ is } \textit{even} \text{ in case } f(-x) = f(x).$$

We see that arctan is odd as follows. Define the function n by $n(x) = -x$; then

$$\arctan(-x) = \int_0^{-x} \frac{1}{1+t^2}\,dt = \int_0^{n(x)} \frac{1}{1+t^2}\,dt = (\arctan \circ n)(x),$$

and by the chain rule,

$$D(\arctan \circ n)(x) = \frac{1}{1+x^2}\,Dn(x) = \frac{-1}{1+x^2} = D(-\arctan x).$$

There is a constant function $\phi(x) = k$, for which

$$\arctan(-x) = (\arctan \circ n)(x) = -\arctan x + \phi(x)$$

and for $x = 0$, $\phi(x) = 0$.

Arctan is bounded. If $0 \leq x \leq 1$,

$$\int_0^x \frac{1}{1+t^2}\,dt \leq x \max\left\{\frac{1}{1+t^2}\,\middle|\, 0 \leq t \leq 1\right\}, \quad 0 \leq x \leq 1.$$

If $1 < x$, then set $\int_0^1 \frac{1}{1+t^2}\,dt = K$, and

$$\int_0^x \frac{1}{1+t^2}\,dt = \int_0^1 \frac{1}{1+t^2}\,dt + \int_1^x \frac{1}{1+t^2}\,dt$$

$$< K + \int_1^x \frac{1}{t^2}\,dt = K + 1 - \frac{1}{x} < K + 1.$$

Therefore, $|\arctan x| < K + 1$.

By the least upper bound axiom, lub range arctan exists.

DEFINITION 7.4 (of π) $\pi = 2$ lub range arctan.

Notice that $\pi/2 \notin$ range arctan, for if it did, then $\int_0^x \frac{1}{1+t^2}\,dt = \pi/2$ for some x, and for $h > 0$, $\arctan(x+h) = \int_0^{x+h} \frac{1}{1+t^2}\,dt = \pi/2 + \int_x^{x+h} \frac{1}{1+t^2}\,dt > \pi/2$, contradicting the definition of π. Since arctan is continuous, every number y such that $|y| < \pi/2$ is in range arctan.

Arctan is strictly monotone increasing, since $D \arctan x = 1(1 + x^2) > 0$, and the mean value theorem implies that for $y > x$,

$$\arctan y - \arctan x = \frac{1}{1+c^2}(y - x) > 0, \quad y < c < x.$$

Therefore, there is a strictly monotone increasing differentiable function inverse to arctan which we temporarily call

$$\text{Tan} = \arctan^{-1}$$

with dom Tan $= \{s \mid -\pi/2 < s < \pi/2\}$, range Tan $=$ {all real numbers} and

$$D\,\text{Tan} = \frac{1}{(D \arctan) \circ \text{Tan}} = 1 + \text{Tan}^2.$$

We proceed to construct the standard trigonometric functions, using this temporary function Tan for a building block.

Suppose that s is not an odd multiple of $\pi/2$. Then $(1/\pi)(s - \pi/2)$ is not an integer, and there is some integer n such that

$$n - 1 < \frac{1}{\pi}\left(s - \frac{\pi}{2}\right) < n.$$

That is,

$$\frac{-\pi}{2} < s - n\pi < \frac{\pi}{2}$$

so that $s - n\pi \in$ dom Tan.

DEFINITION 7.5 The *tangent* function is

$$\tan\colon \tan s = \operatorname{Tan}(s - n\pi), \quad s \text{ not an odd multiple of } \frac{\pi}{2},$$

$$s - \frac{\pi}{2} < n\pi < s + \frac{\pi}{2}, \quad n \text{ an integer.}$$

Note that dom tan is all real numbers except odd multiples of $\pi/2$, and range tan is all real numbers.

A function f is *periodic* with period p if there exists a smallest positive number $p > 0$ such that each $x \in \operatorname{dom} f$

$$f(x+p) = f(x), \quad x \in \operatorname{dom} f.$$

THEOREM 7.4 The function tan is periodic with period π.

Proof: Choose $s \in \operatorname{dom} \tan$; then s is not an odd multiple of $\pi/2$ and neither is $s + \pi$, so $s + \pi \in \operatorname{dom} \tan$ also. For some n

$$s - \frac{\pi}{2} < n\pi < s + \frac{\pi}{2},$$

and therefore

$$(s+\pi) - \frac{\pi}{2} < (n+1)\pi < (s+\pi) + \frac{\pi}{2}$$

giving, by Definition 7.5,

$$\tan s = \operatorname{Tan}(s - n\pi) = \operatorname{Tan}((s+\pi) - (n+1)\pi) = \tan(s+\pi). \ \blacksquare$$

See Figure 7.1.

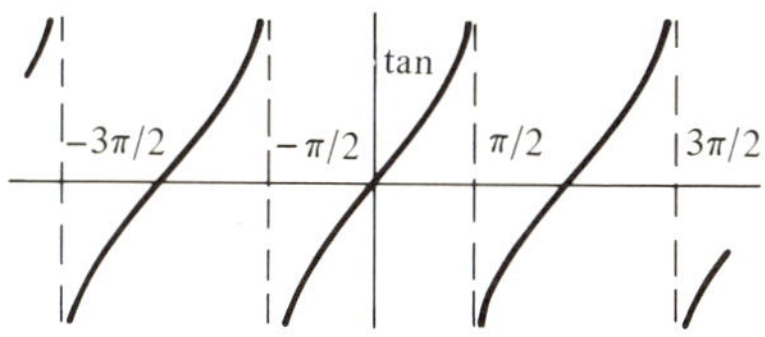

Figure 7.1

THEOREM 7.5 The function tan is differentiable.

Proof:
$$\begin{aligned} D \tan s &= D \operatorname{Tan}(s - n\pi), \quad \text{for some integer } n \\ &= 1 + \operatorname{Tan}^2(s - n\pi) \\ &= 1 + \tan^2 s. \ \blacksquare \end{aligned}$$

Next let us study a particular composition of the function given by the formula $(1 + x^2)^{-1/2}$ with the tan function, namely,

$$f(s) = (1 + \tan^2 s)^{-1/2}, \quad s \text{ not an odd multiple of } \frac{\pi}{2}.$$

Clearly, f is an even function with period π. We endeavor to extend the domain of f to include odd multiples of $\pi/2$, and so to modify f that we obtain a differentiable function on the set of real numbers. We will call this final function *cosine*.

Consider $\pi/2 \notin \operatorname{dom} f$. If we restrict f to $0 \leqslant s < \pi/2$, and choose any ϵ with $0 < \epsilon < 1$, there is an $s > \arctan \sqrt{(1/\epsilon^2) - 1}$ (remember arctan has no maximum value). Since tan is monotone increasing on this interval $[0,\pi/2)$, $\tan s > \sqrt{(1/\epsilon^2) - 1}$, and therefore $0 < f(s) < \epsilon$. Thus, $\lim_{(\pi/2)^-} f = 0$ and with the known properties of tan, $\lim_{(\pi/2)^+} f = 0$ also. Therefore, if we choose

$$f\left((2k + 1)\,\frac{\pi}{2}\right) = 0, \quad k \text{ an integer},$$

we get a continuous function on the real number domain, which we define to be

$$\begin{aligned} {}_1f\colon {}_1f(s) &= (1 + \tan^2 s)^{-1/2}, \quad s \neq (2k + 1)\,\frac{\pi}{2}, \quad k \text{ an integer} \\ &= 0 \qquad\qquad\qquad\quad\;\, , \quad s = (2k + 1)\,\frac{\pi}{2}. \end{aligned}$$

Consider the differentiability of ${}_1f$. If $s \in [0,\pi/2)$, then

$$D_1f(s) = -(1 + \tan^2 s)^{-1/2} \tan s.$$

Before looking for $D_1f(\pi/2)$, we prove the next lemma.

Lemma to Definition 7.6: $\displaystyle\lim_{h \to 0} h \tan\left(\frac{\pi}{2} + h\right) = -1.$

Proof: Apply the double mean value theorem on the interval $(0,h]$ to

$$h \tan\left(\frac{\pi}{2} + h\right) = \frac{h}{\left(\tan\left(\frac{\pi}{2} + h\right)\right)^{-1}}$$

and find

$$h \tan\left(\frac{\pi}{2} + h\right) = \left[1 + \tan^2\left(\frac{\pi}{2} + c\right)\right]^{-1} - 1, \quad 0 < c < h. \ \blacksquare$$

The slope function for ${}_1f$ at $\pi/2$ is

$$
\begin{aligned}
{}_{1f}\sigma_{\pi/2}(h) &= \frac{1}{h}\left[{}_1f\left(\frac{\pi}{2}+h\right) - {}_1f\left(\frac{\pi}{2}\right)\right] \\
&= \frac{1}{h}\left[1+\tan^2\left(\frac{\pi}{2}+h\right)\right]^{-1/2} \\
&= \left[h^2+\left(h\tan\left(\frac{\pi}{2}+h\right)\right)^2\right]^{-1/2}, \quad h>0 \\
&= -\left[h^2+\left(h\tan\left(\frac{\pi}{2}+h\right)\right)^2\right]^{-1/2}, \quad h<0,
\end{aligned}
$$

and so

$$\lim_{0^+} {}_{1f}\sigma_{\pi/2} = 1, \quad \lim_{0^-} {}_{1f}\sigma_{\pi/2} = -1.$$

The function ${}_1f$ is not differentiable at $\pi/2$, and its graph has a corner at $x=(2k+1)\pi/2$, k an integer. See Figure 7.2.

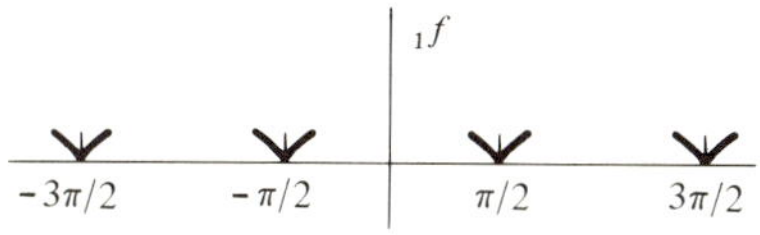

Figure 7.2

However, if we define a function ${}_2f$ by

$$
\begin{aligned}
{}_2f(s) &= {}_1f(s) = (1+\tan^2 s)^{-1/2}, \quad (4k-1)\frac{\pi}{2} \leqslant s \leqslant (4k+1)\frac{\pi}{2} \\
&= -{}_1f(s) = -(1+\tan^2 s)^{-1/2}, \quad (4k+1)\frac{\pi}{2} \leqslant s \leqslant (4k+3)\frac{\pi}{2},
\end{aligned}
$$

for each integer k, we get a differentiable function ${}_2f$ on the domain of real numbers.

DEFINITION 7.6 The function *cosine* is

$$\cos:\ \cos s = {}_2f(s), \quad s \text{ real},$$

where ${}_2f$ is constructed above. See Figure 7.3.

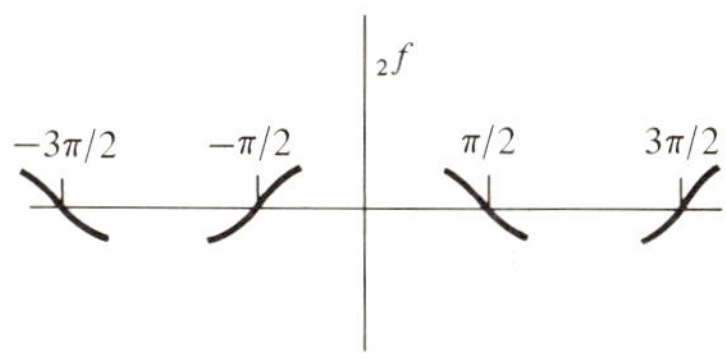

Figure 7.3

DEFINITION 7.7 The function *sine* is

$$\begin{aligned}\sin\colon \sin s &= \sqrt{1-\cos^2 s}\;, \quad 2k\pi \leq s \leq (2k+1)\pi,\\ &= -\sqrt{1-\cos^2 s}, \quad (2k-1)\pi < s < 2k\pi,\end{aligned}$$

where k is any integer.

THEOREM 7.6

(a) $\sin^2 s + \cos^2 s = 1, \quad s$ real

(b) $\tan s = \dfrac{\sin s}{\cos s}, \quad s$ real, $\quad s \neq \dfrac{2k+1}{2}\pi, \quad k$ an integer

(c) $\tan(s+r) = \dfrac{\tan s + \tan r}{1 - \tan s \tan r}, \quad s \in \text{dom tan}, \quad s+r \in \text{dom tan}.$

Proof: Part (a) is immediate from Definition 7.7. A combination of Definitions 7.6 and 7.7 verifies (b). For part (c) we first examine the Tan function, and assume that s, r, and $s+r$ are in $(-\pi/2,\pi/2)$. Set

$$\text{(A)} \qquad \text{Tan}\, s = u, \quad \text{Tan}\, r = v, \quad \text{Tan}(s+r) = w.$$

Then

$$s = \int_0^u \frac{1}{1+t^2}\,dt, \quad r = \int_0^v \frac{1}{1+t^2}\,dt, \quad s+r = \int_0^w \frac{1}{1+t^2}\,dt.$$

Choose any particular number u, and define a function

$${}_u\omega\colon {}_u\omega(v) = w, \quad v \text{ real}, \quad w = \text{Tan}(s+r)$$

with ${}_u\omega(0) = u$. If such a function exists, it must satisfy (see (A))

$$\text{(B)} \qquad \int_0^u \frac{1}{1+t^2}\,dt + \int_0^v \frac{1}{1+t^2}\,dt = \int_0^{{}_u\omega(v)} \frac{1}{1+t^2}\,dt,$$

which, by differentiating line (B), is equivalent to

$$\text{(C)} \qquad \frac{1}{1+v^2} = \frac{D_u\omega(v)}{1+{}_u\omega^2(v)}, \quad {}_u\omega(0) = u.$$

We test[3]

$${}_u\omega(v) = \frac{u+v}{1-uv}$$

and find that it does satisfy line (C). Therefore,

[3] The discovery of this formula is an exercise in solving the special differential equation (C). The method of discovery does not concern us here.

$$\text{Tan}(s + r) = w = {}_u\omega(v) = \frac{\text{Tan } s + \text{Tan } r}{1 - \text{Tan } s \text{ Tan } r}.$$

Using the periodicity of tan, we easily find

$$\tan(s + r) = \frac{\tan s + \tan r}{1 - \tan s \tan r}, \quad s, s + r \in \text{dom tan}$$

also. ■

Corollary to Theorem 7.6

(a) $\sin(s + r) = \sin s \cos r + \cos s \sin r.$
(b) $\cos(s + r) = \cos s \cos r - \sin s \sin r.$

Proof: Verification of these formulas is a routine application of parts (b) and (c) of Theorem 7.6, combined with the definitions of sin and cos. ■

THEOREM 7.7 $D \sin = \cos, \quad D \cos = -\sin.$

Proof: Apply the definitions of sin and cos, and remember that $D \tan = 1 + \text{Tan}^2$. ■

We have completed a basis for analytic trigonometry. In the next chapter we will see how these functions are useful for integration.

7.4 PROBLEMS

1. Sketch graphs of the following.
   (a) arctan (b) Tan
   (c) tan (d) sin
   (e) cos
2. Define the *cotangent, secant,* and *cosecant* functions by

$$\cot = \frac{1}{\tan}, \quad \sec = \frac{1}{\cos}, \quad \csc = \frac{1}{\sin}.$$

   For each of these functions
   (a) specify its domain and range,
   (b) find its period,
   (c) determine whether it is even, odd, or neither,
   (d) examine its continuity,
   (e) examine its differentiability; find a formula for its derivative,
   (f) sketch its graph.
3. Prove that
   (a) $\sec^2 s = 1 + \tan^2 s$;
   (b) $\csc^2 s = 1 + \cot^2 s$.
4. Let k be an integer. List the values of each of the six trigonometric functions at $(k/2)\pi$, whenever $(k/2)\pi$ is in the domain.

5. Proceed as in Problem 4 for the derivatives of each of the six trigonometric functions.
6. Prove each of the following.
   (a) $\cos(\pi/2 - s) = \sin s$
   (b) $\sin(\pi/2 - s) = \cos s$
   (c) $\cos(\pi/2 + s) = -\sin s$
   (d) $\sin(\pi/2 + s) = \cos s$
   (e) $\sin 2s = 2 \sin s \cos s$
   (f) $\cos 2s = 1 - 2 \sin^2 s = 2 \cos^2 s - 1$
7. Prove the following.
   (a) $\sin r \sin s = \frac{1}{2}[\cos(r-s) - \cos(r+s)]$
   (b) $\cos r \cos s = \frac{1}{2}[\cos(r-s) + \cos(r+s)]$
   (c) $\sin r \cos s = \frac{1}{2}[\sin(r-s) + \sin(r+s)]$
8. Prove the following.
   (a) $\sin r + \sin s = 2 \sin \frac{1}{2}(r+s) \cos \frac{1}{2}(r-s)$
   (b) $\cos r + \cos s = 2 \cos \frac{1}{2}(r+s) \cos \frac{1}{2}(r-s)$
9. Prove that the restriction of sine to $[-\pi/2,\pi/2]$ has an inverse. Call this *arcsin*. Sketch its graph and find its derivative.
10. Prove that the restriction of cosine to $[0,\pi]$ has an inverse. Call this *arc cos*. Sketch its graph and find its derivative.
11. Find Df if
    (a) $f(x) = \sin(x^2 + 1)$
    (b) $f(x) = \sin(x^3 + x)$
    (c) $f(x) = \sin(1/x)$
    Answer: $Df(x) = -(\cos(1/x))/x^2$
    (d) $f(y) = \sin(y/(1+y))$
    (e) $f(x) = \cos\sqrt{x}$
    Answer: $Df(x) = -(\sin\sqrt{x})/(2\sqrt{x})$
    (f) $f(t) = \sin(\cos t)$
    (g) $f(z) = \cos(z + \cos z)$
    (h) $f(x) = \sin(1+x) + \cos(1-x)$
    (i) $f(x) = \cos e^x$
    (j) $f(x) = 2\cos(\log_e x) + 3\sin(xe^{x^2})$.
12. Find Df if $f(x) =$
    (a) $\sqrt{1 + \sin x}$

    Answer: $Df(x) = \dfrac{\cos x}{2\sqrt{1+\sin x}}$

    (b) $\sin^2 x$
    Answer: $Df(x) = 2 \sin x \cos x$

    (c) $(\sin x + \cos 2x)^3$
    (d) $(\sin x)/(\cos x)$

    Answer: $Df(x) = \dfrac{1}{\cos^2 x}$

(e) $1/(\sin x \cos x)$

(f) $\dfrac{\sqrt{1+\cos 2x}}{\sin(x/3)}$

(g) $e^{\sin x}$

Answer: $Df(x) = \cos x e^{\sin x}$

(h) $\log_e\left(2 \sin \dfrac{x}{2}\right)$

(i) $1/(\sin x)$

(j) $\cos x^2 \cos x^3$

(k) $(\cos x)/(\sin \sqrt{x})$

(l) $e^{\cos(1/x)} \log_e(\sin^2 4x)$.

13. For each of the following functions find the derivative, and state the domain of the function and of its derivative.

(a) $f(x) = \arcsin 2x$

Answer: $Df(x) = \dfrac{2}{\sqrt{1-4x^2}}$

$-\frac{1}{2} \leqslant x \leqslant \frac{1}{2}$

$-\frac{1}{2} < x < \frac{1}{2}$

(b) $g(x) = \arcsin(x/2)$

(c) $h(x) = \arcsin \dfrac{x^2}{1+x^2}$

(d) $f(y) = \arcsin(1-y^2)$

(e) $g(t) = \arcsin \sqrt{t}$

Answer: $Dg(t) = \dfrac{1}{2\sqrt{t}\,\sqrt{1-t}}$

$0 \leqslant t \leqslant 1$

$0 < t < 1$

(f) $\phi(x) = \arcsin(1 + \cos x)$

14. Draw the graph and find the derivative of f if

(a) $f(x) = \begin{cases} \arcsin x \quad , & -1 \leqslant x < 0 \\ \sin x \quad , & 0 \leqslant x < \pi \\ 1+\cos^2 x, & \pi \leqslant x \leqslant 2\pi \end{cases}$

(b) $f(x) = \begin{cases} \sin 2x \quad , & -\pi \leqslant x < 0 \\ \sin(x/2), & 0 \leqslant x \leqslant 2\pi. \end{cases}$

15. Find $(d^2f)/(dx^2)$ if $f(x) =$

(a) $\cos x$

Answer: $\dfrac{d^2f(x)}{dx^2} = -\cos x$

(b) $\sin 2x$

(c) $x \sin x$

(d) $e^{-x} \sin x$

(e) $e^{2x} \cos\left(\dfrac{x}{2}\right)$.

16. Find dy/dx if
(a) $\cos(x+y) = \sin xy$
Answer: $\dfrac{dy}{dx} = -\dfrac{y \cos xy + \sin(x+y)}{x \cos xy + \sin(x+y)}$
(b) $e^y = \sin x$
(c) $\arcsin xy = y \sin x$.
17. Find equations for the tangent and normal lines to the graphs of the following functions at the indicated points.
(a) $f(x) = x \sin x$, $x = 0$, $x = \pi/4$, and $x = \pi/2$
(b) $g(t) = e^{\sin t}$, $t = 0$, $t = \pi/3$, and $t = 2\pi$
(c) $h(w) = \arcsin(1/w)$, $w = 2$
18. Find the tangent of the angle of intersection of the following pairs of curves.
(a) $y = \sin x$, $y = \cos x$
Answer: $2\sqrt{2}$
(b) $y = x + \sin 4x$, $y = x$
19. Find the extreme values and inflection points of the following functions for the indicated intervals and sketch their graphs. (At an inflection point the second derivative is 0, and there is an open interval on which the second derivative changes sign at that point.)
(a) $f(x) = \sin x + \cos x$, $[0,2\pi]$
Answer: max at $(\pi/4, \sqrt{2})$; min at $(5\pi/4, -\sqrt{2})$; inflection points at $(3\pi/4, 0)$ and $(7\pi/4, 0)$.
(b) $f(x) = x + 2 \sin x$, $[0,2\pi]$
(c) $f(x) = x + \sin 2x$, $[0,2\pi]$
(d) $f(x) = x \sin x$, $[0,2\pi]$
20. A particle moves on a line, its position at any time t being $s(t)$. Investigate its velocity and represent the motion graphically if
(a) $s(t) = 2 \sin 2t$, $0 \leqslant t \leqslant 2\pi$
(b) $s(t) = 2 \sin(t/2)$, $0 \leqslant t \leqslant 2\pi$
(c) $s(t) = \sin t + 2 \cos 2t$, $0 \leqslant t \leqslant \pi$
(d) $s(t) = \sin t \cos 2t$, $0 \leqslant t \leqslant \pi$.
21. Show that $\lim_{x\to 0}(\sin x/x) = 1$.
Hint: For any function f, $Df(a) = \lim_{x\to a} \dfrac{f(x) - f(a)}{x - a}$; also $D \sin 0 = \cos 0 = 1$.
22. Find Df if
(a) $f(x) = \tan(x^2 + 1)$
Answer: $2x \sec^2(x^2 + 1)$
(b) $f(x) = \cot\left(\dfrac{1}{x} + x\right)$
(c) $f(x) = \sec \sqrt{x}$
Answer: $\dfrac{\sec \sqrt{x} \tan \sqrt{x}}{2\sqrt{x}}$

(d) $f(x) = \csc\left(\dfrac{x}{1+x}\right)$
(e) $f(t) = \tan(e^t) \cot(\log t)$
(f) $f(w) = \sec(\sin w) - 2 \sin(\sec w)$
(g) $f(s) = \cos(\csc(\tan(1 - s^2)))$.

23. Find the derivative of each of the following functions.
(a) $\phi(x) = \log \sec(r^2/2)$
Answer: $D\phi(x) = x \tan(x^2/2)$
(b) $\phi(t) = \sqrt{2 \cos(\tan t) + \tan(2 \cos t)}$
(c) $\phi(w) = w^2 \sec^2(w^2) \tan^2(w^2)$
(d) $g(x) = \dfrac{\sec(1-x)}{\sec(1-x) + \tan(1-x)}$
(e) $g(x) = e^{\tan x^2}$
Answer: $Dg(x) = 2x \sec^2 x^2(e^{\tan x^2})$
(f) $g(x) = e^{\tan^2 x}$
(g) $f(x) = \tan(e^{x^2})$
(h) $f(z) = \sec(e^{x^2})e^{\sec x^2}$
(i) $f(x) = \log_e(\sec x + \tan x)$
Answer: $Df(x) = \sec x$

24. Find dy/dx if
(a) $\tan xy = \cot(x + y)$
(b) $\arcsin(x \tan xy) = y \tan x \sec xy$
(c) $e^{\tan(x+y)} = \log_e(\csc xy)$.

25. Find the second derivative of each of the following functions.
(a) $f(x) = \tan^2 x$
Answer: $D^2f(x) = 2 \sec^4 x + 4 \tan^2 x \sec^2 x$
(b) $g(x) = \tan x^2$
(c) $h(x) = e^x \sec x$
(d) $\phi(x) = x \cot x$

26. Find equations for the tangent and normal lines to the graphs of the following functions at the indicated points.
(a) $f(x) = \sin x \tan x$, $x = 0$, $x = \pi/3$, and $x = -\pi/3$
(b) $f(x) = \cot x + \csc(1/x)$, $x = \pi/3$
(c) $f(x) = \log_e(\tan x)$, $x = \pi/4$

27. Find the tangent of the angle of intersection of the following pairs of curves within the indicated intervals.
(a) $y = \sin x$, $y = \tan x$, $[-\pi,\pi]$
Answer: 0
(b) $y = \sin 2x$, $y = \tan(x/2)$, $[-\pi,\pi]$

28. Find the extreme values and inflection points of the following functions, and sketch their graphs for the indicated intervals.
(a) $f(x) = \sin x + \tan x$, $[0,\pi]$
(b) $f(x) = 2x + \tan x$, $[0,\pi]$
Answer: No extreme values. Inflection point at (0,0).

(c) $f(x) = x \tan x, \quad [0,\pi]$

29. Find Df if $f(x) =$
   (a) arc cos $4x$

   Answer: $Df(x) = \dfrac{-4}{\sqrt{1-16x^2}}$

   (b) $\arctan(x/4)$
   (c) arc cos$(1+x)$
   (d) arc cos$(1-x^2)$

   Answer: $Df(x) = \dfrac{2}{\sqrt{2-x^2}}$

   (e) $\arctan e^x$
   (f) $\arctan(\sec x + \tan x)$
   (g) arc cos$(\log_e(1+x))$.
30. Find the derivative of each of the following functions.
   (a) $f(x) = \sqrt{\text{arc cos } 2x}$

   Answer: $Df(x) = \dfrac{-1}{\sqrt{\text{arc cos } 2x}\,\sqrt{1-4x^2}}$

   (b) $g(y) = y^2 \arctan y^2$
   (c) $\phi(t) = (\arcsin t)(\text{arc cos } t)$
   (d) $f(t) = \dfrac{t}{\arctan 2t}$
   (e) $h(w) = \sin(\arctan(1/w))$

   Answer: $Dh(w) = \dfrac{-\cos(\arctan(1/w))}{w^2+1}$

   (f) $g(s) = e^{\text{arc cos } 2s}$
31. Find dy/dx if
   (a) arc cos $xy = \arcsin(x+y)$
   (b) $\arctan(x/y) = \log(x+y)$
   (c) x arc cos $y + y \arctan x = \arcsin(x+y)$.

7.5 ELEMENTARY FUNCTIONS

Certain functions have been designated as *elementary*, and we have seen many of them among the examples. This classification is an arbitrary one, and it is not precisely defined in the literature. However, the following functions are usually called elementary.

(a) Polynomial functions
(b) Rational functions
(c) Root functions (with the formula $f(x) = \sqrt[p]{x}$, $p > 0$ a positive integer)
(d) Trigonometric functions
(e) The exponential function
(f) Inverses of (d) and (e)

(g) Functions constructed from the above types by the arithmetic operations and composition

Restrictions of these functions to subdomains are usually included in the elementary class. We note that the elementary functions have elementary derivatives, but not necessarily elementary integrals. For example, $\int \sqrt{\sin x}\, dx$ is not elementary, although it is difficult to prove this.

Chapter 8
INTEGRATION

8.1 THE ASSOCIATED DIFFERENTIAL EQUATION

At the beginning of Chapter 7 we studied the computation of $\int_a^b f$ by an application of the fundamental theorem (Chapter 6, Section 4). This process led to a search for a function g whose derivative is f, and with which we are reasonably familiar. Often this search is not easy. Then, too, some of us know more about functions than others do, so a function which is familiar to one student may be unusable by another. In the last chapter we assumed that log, exp, and the trigonometric functions and their inverses were not familiar to us, and we used the definite integral to invent them. In doing this we were illustrating the following theorem.

THEOREM 8.1 Let f be a continuous function, $\operatorname{dom} f = [a,b]$. Then the function g

$$g(x) = \int_a^x f, \quad a \leqslant x \leqslant b$$

satisfies the equations

$$\text{(A)} \qquad Dg = f, \qquad g(a) = 0,$$

and g is the only function that does.

Proof: Since the restriction of f to the interval $[a,x]$ is continuous, $g(x)$ exists for each x, and g is a function with dom $g = [a,b]$. Finally, by Lemma 5 to Theorem 6.1, $Dg = f$, and $g(a) = \int_a^a f = 0$, so g satisfies equations (A).

Suppose that $\bar{g}$ also represents a function satisfying Equations (A). Then

$$D(g - \bar{g}) = Dg - D\bar{g} = f - f = \underline{0},$$

and so for some number[1] k, $g - \bar{g} = \underline{k}$. But

$$0 = g(a) - \bar{g}(a) = \underline{k}(a),$$

and hence $k = 0$; that is, $\bar{g} = g$. ■

The equation $Dg = f$ is an instance of a *differential equation;* that is, an equation in which a derivative of the "unknown" function is present, and the equation $g(a) = 0$ is called an *initial condition* on a solution of the differential equation. The theorem states that the pair of Equations (A) has a unique solution, and it is provided by the integral formula $\int_a^x f$. From the point of view of solving the differential equation, it is very useful to know this. On the other hand, the theorem also says that if we know a solution of (A) with a reasonable degree of familiarity, then this solution enables us to evaluate the integral $\int_a^b f$. This point of view is our immediate concern.

When we attempted to evaluate an integral such as $\int_2^{10} \frac{1}{t}\, dt$, and could find no rational function whose derivative is the reciprocal function, r, we set

$$g(x) = \int_1^x \frac{1}{t}\, dt = \int_1^x r$$

and proved that $Dg = r$, $g(1) = 0$. Then, for instance,

[1] Recall that if k is a number, then, $\underline{k}$ is the constant function whose range is the singleton set $\{k\}$.

$$\int_2^{10} \frac{1}{t}\, dt = g(10) - g(2) = \int_1^{10} \frac{1}{t}\, dt - \int_1^{2} \frac{1}{t}\, dt.$$

When we write

$$\int_2^{10} \frac{1}{t}\, dt = \log 10 - \log 2 = \log 5$$

we are only giving g the name "log" and using properties of this particular function g which have become familiar after they have been ascertained.

In a similar way

$$\int_1^{\sqrt{3}} \frac{1}{1+t^2}\, dt = g(\sqrt{3}) - g(1) = \int_0^{\sqrt{3}} \frac{1}{1+t^2}\, dt - \int_0^{1} \frac{1}{1+t^2}\, dt,$$

where

$$g(x) = \int_0^{x} \frac{1}{1+t^2}\, dt, \quad g(0) = 0.$$

This is usually written

$$\int_1^{\sqrt{3}} \frac{1}{1+t^2}\, dt = \arctan\sqrt{3} - \arctan 1 = \frac{\pi}{3} - \frac{\pi}{4} = \frac{\pi}{12}.$$

One more recollection will serve to launch us into the subject of this chapter. In Section 6.5 we faced the calculation of the integral

$$\int_0^{r} \sqrt{r^2 - x^2}\, dx, \quad r \text{ is a number.}$$

We know, Theorem 8.1, that there is a function g

$$g(x) = \int_0^{x} \sqrt{r^2 - t^2}\, dt, \quad 0 \leqslant x \leqslant r$$

whose derivative is the function to be integrated.[2] Thus,

$$\int_0^{r} \sqrt{r^2 - x^2}\, dx = g(r) - g(0),$$

and the question is whether this function g is familiar enough to be of use. We observed that it is, namely, the algebraic combination of familiar functions

$$g(x) = \frac{r^2}{2} \arcsin \frac{x}{r} + \frac{x}{2} \sqrt{r^2 - x^2}, \quad 0 \leqslant x \leqslant r.$$

Although this fact can sufficiently be verified by computing Dg and $g(0)$, there is as yet no hint as to how it could have been discovered.

[2] A function to be integrated is called the *integrand.*

Returning to the Proof of Theorem 8.1, remember that we demonstrated the equivalence of the function whose formula is $\int_a^x f$, with the solution g of the equations $Dg = f$, $g(a) = 0$. However, in the application of Theorem 8.1 to the computation of an integral $\int_a^b f$ by the fundamental theorem, we need not have used just g itself, but any function given by $\bar{g} = g + \underline{k}$, k a number. Specifically,

$$\begin{aligned}\int_a^b f &= g(b) - g(a) = (\bar{g} - \underline{k})(b) - (\bar{g} - \underline{k})(a)\\ &= \bar{g}(b) - \underline{k}(b) - \bar{g}(a) + \underline{k}(a)\\ &= \bar{g}(b) - k - \bar{g}(a) + k = \bar{g}(b) - \bar{g}(a).\end{aligned}$$

It is customary to call any such function $\bar{g}$ an *indefinite integral* of f and to represent it by the symbol $\int f(x)dx$. Thus the indefinite integral symbol represents any one of many functions, and each is sufficient to "do" an integration problem; that is, to *integrate* a function. In this sense two functions represented by $\int f(x)dx$ are equivalent.

EXAMPLE 1 Integrate $\int_2^{10} \frac{1}{t}\, dt$.

(a) Take $\bar{g}(x) = \int \frac{1}{x}\, dx = \log x$,

then $\int_2^{10} \frac{1}{t}\, dt = \log 10 - \log 2 = \log 5$.

(b) Take $\bar{g}(x) = \int \frac{1}{x}\, dx = \log x + 8$,

then $\int_2^{10} \frac{1}{t}\, dt = (\log 10 + 8) - (\log 2 + 8) = \log 5$.

EXAMPLE 2 Integrate $\int_1^{\sqrt{3}} \frac{1}{1+t^2}\, dt$.

(a) Take $\bar{g}(x) = \int \frac{1}{1+x^2}\, dx = \arctan x$,

then $\int_1^{\sqrt{3}} \frac{1}{1+t^2}\, dt = \arctan \sqrt{3} - \arctan 1 = \frac{\pi}{12}$.

(b) Take $\bar{g}(x) = \int \frac{1}{1+x^2}\, dx = \arctan x - 7$,

then $\int_1^{\sqrt{3}} \frac{1}{1+t^2}\, dt = (\arctan \sqrt{3} - 7) - (\arctan 1 - 7) = \frac{\pi}{12}$.

EXAMPLE 3 Integrate $\int_0^r \sqrt{r^2 - t^2}\, dt$.

(a) Take $\bar{g}(x) = \int \sqrt{r^2 - x^2}\, dx = \frac{r^2}{2} \arcsin \frac{x}{r} + \frac{x}{2} \sqrt{r^2 - x^2}$,

then[3] $\int_0^r \sqrt{r^2 - t^2}\, dt = \bar{g}(r) - \bar{g}(0) = \frac{\pi r^2}{4}$.

(b) Take $\bar{g}(x) = \int \sqrt{r^2 - x^2}\, dx = \frac{r^2}{2} \arcsin \frac{x}{r} + \frac{x}{2} \sqrt{r^2 - x^2} + 53$,

then $\int_0^r \sqrt{r^2 - t^2}\, dt = \bar{g}(r) - \bar{g}(0) = \frac{\pi r^2}{4}$.

In Example 1 the direction to find $\bar{g}$, that is, to find some $\int (1/x)\, dx$, is written "integrate $1/x$." The answer is $\log x + k$ for some k (any k will do). In Example 2 "integrate $1/(1 + x^2)$" has the answer $\arctan x + k$ for some k (any k will do). Example 3 has a similar characterization. In most such formulas $k = 0$ gives the simplest description of an indefinite integral.

8.1 PROBLEMS

1. Find indefinite integrals for each of the following.

(a) $\int 6x^3\, dx$

Answer: $\frac{3}{2}x^4 + c$

(b) $\int (4x^4 - 7x)\, dx$

(c) $\int (x^{10} - x^9 + x^3 - 10x + 9)\, dx$

(d) $\int \sqrt{x}\, dx$

Answer: $\frac{2}{3}x^{3/2} + c$

(e) $\int x^{2/3} dx$.

(f) $\int (x^{3/2} + x^{-3/2})\, dx$

(g) $\int (1 + \sqrt{x})^2 dx$

[3] It is clearer, in this line, to omit the bloody formula for $\bar{g}$, because you surely see how it works.

(h) $\int \frac{dt}{\sqrt{t}}$

Answer: $2\sqrt{t} + c$

(i) $\int \frac{ds}{s^3}$

(j) $\int \left(x^2 - \frac{1}{x^2}\right) dx$

Answer: $\frac{x^3}{3} + \frac{1}{x} + c$

(k) $\int (x^2 - 4)^2 dx$

Answer: $\frac{x^5}{5} - \frac{8}{3} x^3 + 16x + c$

(l) $\int (x + 1)(x^3 - 3) dx$

(m) $\int (1 + \sqrt{x})^3 dx$

(n) $\int \sqrt{x}(1 + x)^2 dx$

(o) $\int t^{2/3}(1 + t^{-2/3})^2 dt$

(p) $\int y^3(2 - y^2)^3 dy$

(q) $\int \frac{x^3 - 1}{x^2} dx$

Answer: $\frac{x^2}{2} + \frac{1}{x} + c$

(r) $\int \frac{x^4 + x - 3}{x^3} dx$

(s) $\int \frac{(\sqrt{y} + 1)^2}{\sqrt{y}} dy$

(t) $\int (s - 2)^3(2s)^3 ds$

(u) $\int (\sqrt{x} - 1)(\sqrt{x} + 1) dx$

(v) $\int \frac{x - 1}{\sqrt{x} - 1} dx$

(w) $\int (\sqrt{x} + y)\,dx$

(x) $\int (\sqrt{x} + y)\,dy$

2. See Section 7.2, Problem 11, page 209. For part (a) recall that

$$D \log(1 - x^3) = \frac{3x^2}{x^3 - 1},$$

and therefore

$$\int \frac{3x^2}{x^3 - 1}\,dx = \log(1 - x^3) + k$$

for any number k. Write each of the parts (b) through (k) in this way as an indefinite integral.

3. See Section 7.2, Problem 19, page 210. Write each of the parts (a) through (k) as an indefinite integral.
4. Use Problems 2 and 3 to find

(a) $\int_0^1 \frac{x}{x^2 - 4}\,dx$

(b) $\int_e^{2e} \left(\frac{2}{x} + \frac{1}{x \log x}\right) dx, \quad \log e = 1$

(c) $\int_0^1 2xe^{-x^2}dx$

(d) $\int_{1/2}^1 (2w + 1)e^{2w}dw.$

5. Refer to Section 7.4, Problem 11, page 224. Write each of the parts (a) through (j) as an indefinite integral.
6. Use Problem 5 to find

(a) $\int_{2/\pi}^{4/\pi} \frac{1}{x^2} \cos \frac{1}{x}\,dx$

(b) $\int_{\pi/2}^{\pi} \sin t \cos (\cos t)\,dt$

(c) $\int_{\log \pi}^{\log 2\pi} e^x \sin e^x dx.$

8.2 INTEGRATION BY SUBSTITUTION

The problem of computing an integral $\int_a^b f$ by an application of the fundamental theorem,

$$\int_a^b f = g(b) - g(a)$$

is now reduced to solving the equation[4]

(B) $$Dg = f$$

for some g, and any such g is written

$$g(x) = \int f(x)dx.$$

In the happiest case we just remember some g whose derivative, we know, is f. Such simple cases are rare. The following theorem suggests one way to find another integral, say $\int_c^d h$, which is equal to $\int_a^b f$, and for which an indefinite integral, $\int h(x)dx$ may more readily be discovered.

THEOREM 8.2 (transformation of the integral) Let s be a function with a continuous derivative Ds on the interval $[c,d]$, and let f be continuous on the interval $[s(c),s(d)]$, or on $[s(d),s(c)]$. Then

(C) $$\int_c^d (f \circ s)Ds = \int_{s(c)}^{s(d)} f.$$

Proof: Under the hypotheses, both integrals exist; we need only show their equality. By Theorem 8.1 there is a function g such that $Dg = f$. Therefore, by the fundamental theorem and the chain rule,

$$\begin{aligned}\int_c^d (f \circ s)Ds &= \int_c^d [(Dg) \circ s]Ds = \int_c^d D(g \circ s) \\ &= (g \circ s)(d) - (g \circ s)(c) = g(s(d)) - g(s(c)) \\ &= \int_{s(c)}^{s(d)} Dg = \int_{s(c)}^{s(d)} f. \blacksquare\end{aligned}$$

Corollary to Theorem 8.2: Under the hypotheses of Theorem 2,

(D) $$\int_a^b f = \int_{s^{-1}(a)}^{s^{-1}(b)} (f \circ s)Ds.$$

Proof: Set $s(c) = a, \quad s(d) = b$. $\blacksquare$

Consider Formula (C). Suppose, in a given integral, the integrand can be recognized as having the form of the left member of the equation, with the functions f and s identified so that an indefinite integral of f is known. Then the right member of the equation can be integrated by the funda-

[4] We drop the notation $\bar{g}$, since all we need is any solution of (B); that is, any value of $\int f(x)dx$.

mental theorem. When this recognition occurs the function $(f \circ s)\,Ds$ is called an *exact differential.*

EXAMPLE 1 $\displaystyle\int_{\pi/4}^{\pi/2} \frac{\cos t}{\sin t}\,dt$

Set $s(t) = \sin t$; then $Ds(t) = \cos t$, and take $f(x) = 1/x$. Then the integrand can be seen to have the form

$$(f \circ s)Ds,$$

where $s(\pi/4) = 1/\sqrt{2}$, $s(\pi/2) = 1$, and Formula (C) gives

$$\int_{\pi/4}^{\pi/2} \frac{\cos t}{\sin t}\,dt = \int_{1/\sqrt{2}}^{1} \frac{1}{x}\,dx = \log 1 - \log \frac{1}{\sqrt{2}} = \frac{1}{2} \log 2.$$

The variable notation is handy for such integrations. An obvious shorthand would be to write

$$x = \sin t, \qquad dx = \cos t dt$$

$$t = \frac{\pi}{4} \Longrightarrow x = \frac{1}{\sqrt{2}}, \quad t = \frac{\pi}{2} \Longrightarrow x = 1$$

and then just translate the t variable to the x variable.

This problem can also be solved by explicit use of the indefinite integrals, as follows. The integral $\displaystyle\int_{\pi/4}^{\pi/2} \frac{\cos t}{\sin t}\,dt$ could be evaluated if we knew some indefinite integral $\displaystyle\int \frac{\cos t}{\sin t}\,dt$. When we identify the functions f and s just as before, we find

$$\int \frac{\cos t}{\sin t}\,dt = \int \frac{1}{\sin t} \cos t\,dt = \int \frac{1}{x}\,dx = \log x + k = \log \sin t + k$$

for some number k (any k will do). With $k = 0$ we have an indefinite integral $\log \sin t$ for the original integral, and

$$\int_{\pi/4}^{\pi/2} \frac{\cos t}{\sin t}\,dt = \log \sin \frac{\pi}{2} - \log \sin \frac{\pi}{4} = \frac{1}{2} \log 2.$$

EXAMPLE 2 $\displaystyle\int_0^1 t(1 + t^2)^4 dt$

The binomial can be expanded to five terms, and the integral evaluated easily since polynomials are familiar functions. However, on the pattern of Formula (C), if we take

$$s(t) = 1 + t^2, \qquad f(x) = x^4,$$

then $(f \circ s)(t) = (1 + t^2)^4$. But $Ds(t) = 2t$, and we missed the integrand by a factor 2. However, this can be corrected as follows:

$$\int_0^1 t(1+t^2)^4 dt = \frac{1}{2}\int_0^1 (1+t^2)^4 2t\, dt = \frac{1}{2}\int_1^2 x^4\, dx = \frac{1}{10}[2^5 - 1^5] = \frac{31}{10}.$$

Alternatively, we could do this with the variables alone. If $x = 1 + t^2$, $dx = 2t\, dt$. Then, using indefinite integrals

$$\int t(1+t^2)^4 dt = \frac{1}{2}\int x^4 dx = \frac{1}{10}x^5 = \frac{1}{10}(1+t^2)^5$$

and so

$$\int_0^1 t(1+t^2)^4 dt = \frac{1}{10}(1+1^2)^5 - \frac{1}{10}(1+0^2)^5 = \frac{32}{10} - \frac{1}{10} = \frac{31}{10}.$$

Notice that in the variable shorthand we lose sight of the functions involved and of the theorem which justifies the process. With practice the shorthand does become very fast, efficient, and with care, accurate.

8.2 PROBLEMS

1. Identify f and s on the left side of Formula (C). Use the right side to find the following indefinite integrals.

(a) $\int \sqrt{2x}\, dx$

Answer: $\frac{2\sqrt{2}}{3}x^{3/2} + c$

(b) $\int \sqrt{\frac{x}{2}}\, dx$

(c) $\int x\sqrt{1 - x^2}\, dx$

(d) $\int y^3\sqrt{4 + y^4}\, dy$

(e) $\int \frac{t}{\sqrt[3]{t^2 - 1}}\, dt$

Answer: $\frac{3}{4}(t^2 - 1)^{2/3} + c$

(f) $\int \frac{s^4}{(1 - 2s^5)^{1/5}}\, ds$

(g) $\int \frac{6x^2 - 2}{\sqrt{x^3 - x}}\, dx$

(h) $\int x^2(2 - 3x^3)^4\, dx$

(i) $\int \sqrt{1 - w}\, dw$

Answer: $-\frac{2}{3}(1 - w)^{3/2} + c$

(j) $\int \sqrt{2 - 3w}\, dw$

(k) $\int (3x^2 - 4x^3)^{4/5}(2x - 4x^2)\, dx$

(l) $\int \frac{\sqrt{1 + \sqrt{x}}}{\sqrt{x}}\, dx$

2. Solve the differential equation $Dg = f$ for g, where f is given, in each of the following cases.

(a) $f(x) = \sqrt[4]{4 - x}$

(b) $f(x) = x^3 \sqrt[4]{4 - x^4}$

(c) $f(z) = \left(z + \frac{1}{z}\right)^3$

(d) $f(t) = \frac{t}{(1 - t^2)^3}$

(e) $f(s) = \frac{s - 1}{(s^2 - 2s + 3)^2}$

(f) $f(y) = \frac{(2 - y^{1/3})^2}{y^{1/3}}$

3. Find the following indefinite integrals.

(a) $\int 2 \sin 2x\, dx$

(b) $\int \sin 2x\, dx$

(c) $\int \sin wx\, dx$

(d) $\int 2x \sin x^2\, dx$

(e) $\int x \sin x^2\, dx$

4. Find the indefinite integrals, n is a positive integer,

(a) $\int x^{n-1} \sin x^n\, dx$

(b) $\int x^{n-1} \cos x^n\, dx$

(c) $\int x^{n-1} e^{x^n}\, dx$

(d) $\int x^{n-1} \sec^2 x^n\, dx.$

5. Find the indefinite integrals, n is an integer,

(a) $\int \sin^2 x \cos x\, dx$

(b) $\int \cos^2 x \sin x\, dx$

(c) $\int \sin^n x \cos x\, dx, \quad n = -1$ is special

(d) $\int \cos^n x \sin x\, dx$

(e) $\int \sin^2 x \cos^3 x\, dx, \quad \cos^3 x = (1 - \sin^2 x) \cos x$

(f) $\int \sin^5 x \cos^4 x\, dx.$

6. Find

(a) $\int \frac{1}{x} \sqrt{\log x}\, dx$ (b) $\int \frac{1}{x} \log^n x\, dx.$

Now consider Formula (D) in the Corollary to Theorem 8.2. Suppose that the integral $\int_a^b f$ is to be evaluated and that no indefinite integral $\int f(x)dx$ is apparent. It may be possible to discover some function s so that an indefinite integral is available for the equivalent integral

$$\int_{s^{-1}(a)}^{s^{-1}(b)} (f \circ s) Ds.$$

When this is done, the integral $\int_a^b f$ is said to have been *integrated by substitution*. The function s is the function *substituted*.

EXAMPLE 3 $\int_{-1}^{0} x\sqrt{1 - x}\, dx$

In this case the function f is

$$f(x) = x\sqrt{1 - x}, \quad -1 \leqslant x \leqslant 0.$$

Choose the substitution function, $s(t) = 1 - t^2$, $1 \leqslant t \leqslant \sqrt{2}$, and we will explain shortly why this is a plausible choice for s. Given this, however, $Ds(t) = -2t$ and the Formula (D) gives

$$\begin{aligned}\int_{-1}^{0} x\sqrt{1 - x}\, dx &= \int_{\sqrt{2}}^{1} (1 - t^2)\, \sqrt{1 - (1 - t^2)}\, (-2t)\, dt \\ &= -2 \int_{\sqrt{2}}^{1} (1 - t^2) t^2\, dt \\ &= 2 \int_{1}^{\sqrt{2}} (t^2 - t^4) dt = -\frac{4}{15} (1 + \sqrt{2}),\end{aligned}$$

in which you can see the composition at work, and the result is easy to finish.

Choosing s pulled off the trick. In forming the composition we set $x = s(t) = 1 - t^2$, and this is equivalent to $t = s^{-1}(x) = \sqrt{1 - x}$; that is, we chose s so that the root function would, in the composition, be replaced by the simpler identity function. Let us repeat this computation by the indefinite integral method.

Consider $\int_{-1}^{0} x\sqrt{1-x}\, dx$. Set $\sqrt{1-x} = t$ as a hopeful start, only because the root function seems troublesome. Since $-1 \leqslant x \leqslant 0$, then $1 \leqslant t \leqslant \sqrt{2}$ and, solving for x, $x = 1 - t^2$. This hopeful start has produced the substitution function $s(t) = 1 - t^2$. Then, as before, $Ds(t) = -2t$, and by the pattern of Formula (D),

$$\begin{aligned}\int x\sqrt{1-x}\, dx &= \int (1-t^2)t\,(-2t)\,dt \\ &= -2\int (t^2 - t^4)\,dt \\ &= -\frac{2t^3}{15}(5-3t^2) \\ &= -\frac{2}{15}(1-x)^{3/2}\,[5-3(1-x)] \\ &= -\frac{2}{15}(1-x)^{3/2}\,[2+3x].\end{aligned}$$

This gives an indefinite integral for the given integral. Thus,

$$\int_{-1}^{0} x\sqrt{1-x}\, dx = -\frac{2}{15}[2+2\sqrt{2}] = -\frac{4}{15}(1+\sqrt{2}).$$

EXAMPLE 4 $\displaystyle\int_{-1}^{1} \frac{1}{\sqrt{4-x^2}}\, dx$

As in the previous example it is the root function that introduces difficulty. If you search for a helpful substitution function by the trial $\sqrt{4-x^2} = t$ you will discover that the root function recurs in the right side of Formula (D), and there has been no gain. In this case we try the substitution function $x = s(t) = 2 \sin t$, $-\pi/6 \leqslant t \leqslant \pi/6$, because then $\sqrt{4-x^2} = \sqrt{4 - 4\sin^2 t} = 2\cos t$. Since $Ds(t) = 2\cos t$,

$$\int_{-1}^{1} \frac{1}{\sqrt{4-x^2}}\, dx = \int_{-\pi/6}^{\pi/6} 1\,dt = \frac{\pi}{3}.$$

You should repeat this example using the substitution function $s(t) = 2\cos t$, $\pi/3 \leqslant t \leqslant 2\pi/3$; and again by the indefinite integral method.

8.2 PROBLEMS (*continued*)

7. Define a transformation of the integral $\int_a^b f$ by the substitution s to be

$$T_s \int_a^b f = \int_{s^{-1}(a)}^{s^{-1}(b)} (f \circ s)Ds.$$

Then the Corollary to Theorem 8.2 states that $T_s \int_a^b f = \int_a^b f$. Notice that the integrand and the limits in $T_s \int_a^b f$ are generally different from those in $\int_a^b f$. Show that

$$T_{s^{-1}}\, T_s \int_a^b f = \int_a^b f,$$

where the integrand the limits in both $T_{s^{-1}}\, T_s \int_a^b f$ and $\int_a^b f$ are the same.

8. Find the following indefinite integrals.

(a) $\int x^2\sqrt{x-1}\,dx$ (b) $\int x\sqrt{1+2x}\,dx$

(c) $\int x\sqrt{3x-1}\,dx$ (d) $\int \frac{y\,dy}{\sqrt{1+y}}$

(e) $\int \frac{s^2\,ds}{\sqrt{1-3s}}$ (f) $\int x^3\sqrt{10-(x/2)}\,dx$

(g) $\int \frac{t\,dt}{\sqrt{2-3t}}$

9. (a) $\int x\sqrt[3]{1+x}\,dx$

Answer: $\frac{3}{7}(1+x)^{7/3} - \frac{3}{4}(1+x)^{4/3} + c$

(b) $\int x^2\sqrt[3]{2x-3}\,dx$ (c) $\int y\sqrt[4]{1+y}\,dy$

(d) $\int s(1+s)^{2/3}\,ds$ (e) $\int t(10-5t)^{1/3}\,dt$

(f) $\int \frac{w\,dw}{\sqrt[3]{w-1}}$

10. Find a substitution to integrate each of the following. Check using Theorem 8.1.

(a) $\int (6x)^3\,dx;\quad t = 6x$

(b) $\int x\sqrt{1+x^2}\,dx;\quad t = 1+x^2$

(c) $\int x\sqrt{1+x^2}\,dx;\quad t = \sqrt{1+x^2}$

(d) $\int \sqrt{1-x}\, dx; \quad t = 1 - x$

11. Find

(a) $\int \frac{\sqrt{4x^2 - 1}}{x}\, dx$ (Try $x = \frac{1}{2} \sec t$.)

(b) $\int x^3\sqrt{1-x^2}\, dx$ (Try $x = \cos t$; also try $x = \sin t$.)

(c) $\int \frac{\sqrt{9 - 4x^2}}{x}\, dx$

(d) $\int \sqrt{16 - x^2}\, dx$

(e) $\int x^3\sqrt{x^2 + 4}\, dx.$

12. Find

(a) $\int e^{2x}\, dx$

(b) $\int xe^{x^2}\, dx$

(c) $\int 10^x\, dx$

(d) $\int (e^x - e^{-x})^2\, dx$

(e) $\int (x^2 + 1)e^{x^3+3x}\, dx$

(f) $\int \frac{e^y}{e^y + 2}\, dy$

(g) $\int \frac{3e^{\ln x}}{x}\, dx$

(h) $\int \frac{2^{\sqrt{x}}}{\sqrt{x}}\, dx.$

13. Evaluate the following.

(a) $\int_0^2 e^{x/2}\, dx$

Answer: $2e - 2$

(b) $\int_0^1 \frac{dx}{e^x}$

Answer: $1 - 1/e$

(c) $\int_0^1 \frac{e^t + e^{-t}}{e^t - e^{-t}}\, dt$

(d) $\int_1^2 2^{1-y}\, dy$

8.3 INTEGRATION BY PARTS

We discuss here another way to replace an integral $\int_a^b f$ by a different integral which may be easier to evaluate.

THEOREM 8.3 Let u and v be functions with continuous derivatives on the interval $[a,b]$. Then

$$\text{(E)} \qquad \int_a^b uDv = u(b)v(b) - u(a)v(a) - \int_a^b vDu.$$

Proof: Under the hypotheses each term in the Equation (E) is a defined number. We need only show the equality. By the product formula for derivatives,

$$D(uv) = uDv + vDu,$$

therefore

$$\text{(F)} \qquad \int_a^b D(uv) = \int_a^b uDv + \int_a^b vDu.$$

But the integrand on the left is the derivative of uv; hence, an indefinite integral is $g = uv$, and

$$\int_a^b D(uv) = g(b) - g(a) - u(b)v(b) - u(a)v(a),$$

which, with (F), gives (E). ■

THEOREM 8.4 Under the hypotheses of Theorem 3 any two functions g and $\bar{g}$, where[5]

$$g(x) = \int u(x)Dv(x)dx, \qquad \bar{g}(x) = u(x)v(x) - \int v(x)Du(x)dx,$$

are equivalent indefinite integrals; that is,

$$g = \bar{g} + \underline{k} \text{ for some number } k.$$

Proof: By the definition of an indefinite integral,

$$Dg = uDv, \quad D\bar{g} = D(uv) - vDu,$$

and therefore

$$D(g - \bar{g}) = uDv - D(uv) + vDu = \underline{0}$$

so that $g = \bar{g} + k$ for some number k. ■

[5] Given some pair of indefinite integrals, $\int u(x)Dv(x)dx$ and $\int v(x)Du(x)dx$, then g and $\bar{g}$ are specified functions.

EXAMPLE 5 $\displaystyle\int_0^{\log 5} xe^x dx$

The integrand is the product of two functions and clearly has the form of the left member of Formula (E). The trick is to anticipate how the integrand on the right side of (E) will appear when an identification with the u and v of Theorem 8.3 is made. If we identify

$$u(x) = x, \qquad Dv(x) = e^x,$$

then

$$Du(x) = 1,$$

and we can choose $v(x) = e^x$ so that

$$\begin{aligned}\int_0^{\log 5} xe^x dx = \int_0^{\log 5} xDe^x dx &= \log 5\ e^{\log 5} - \int_0^{\log 5} e^x dx \\ &= 5 \log 5 - [e^{\log 5} - 1] = 5 \log 5 - 4.\end{aligned}$$

Notice that the choice of v is arbitrary to within a constant, any function whose derivative is exp will do. Thus, using indefinite integrals, we again identify

$$u(x) = x, \quad Dv(x) = e^x.$$

Then, in general,

$$Du(x) = 1, \quad v(x) = e^x + k$$

(but we might as well take $k = 0$), so that by Theorem 8.4 an indefinite integral for $\displaystyle\int_0^{\log 5} xe^x dx$ is

$$xe^x - \int e^x dx = xe^x - e^x = e^x(x - 1)$$

and

$$\int_0^{\log 5} xe^x dx = e^{\log 5}(\log 5 - 1) + e^0 = 5 \log 5 - 4.$$

EXAMPLE 6 $\displaystyle\int_0^{\pi/2} e^x \cos x\, dx$

(a) Apply Theorem 8.3. Set

$$u(x) = e^x, \quad Dv(x) = \cos x.$$

Then

$$Du(x) = e^x, \quad v(x) = \sin x + k,$$

and, with $k = 0$,

$$\text{(G)}\quad \begin{aligned}\int_0^{\pi/2} e^x \cos x\, dx &= e^{\pi/2} \sin \frac{\pi}{2} - e^0 \sin 0 - \int_0^{\pi/2} e^x \sin x\, dx \\ &= e^{\pi/2} - \int_0^{\pi/2} e^x \sin x\, dx.\end{aligned}$$

Now treat the new integral on the right in the same way. Set

$$u(x) = e^x, \quad Dv(x) = \sin x.$$

Then

$$Du(x) = e^x, \quad v(x) = -\cos x + k,$$

and, with $k = 0$,

$$\begin{aligned}\int_0^{\pi/2} e^x \sin x\, dx &= -e^{\pi/2} \cos \frac{\pi}{2} + e^0 \cos 0 + \int_0^{\pi/2} e^x \cos x\, dx \\ &= 1 + \int_0^{\pi/2} e^x \cos x\, dx.\end{aligned}$$

Therefore, from (G)

$$\int_0^{\pi/2} e^x \cos x\, dx = e^{\pi/2} - 1 - \int_0^{\pi/2} e^x \cos x\, dx,$$

and so

$$\int_0^{\pi/2} e^x \cos x\, dx = \frac{1}{2}\,(e^{\pi/2} - 1).$$

(b) Apply Theorem 8.4, using the same identification with u and v as in part (a). If an indefinite integral is

$$g(x) = \int e^x \cos x\, dx,$$

then by Theorem 8.4, so is

$$\bar{g}(x) = e^x \sin x - \int e^x \sin x\, dx,$$

where the second term is also an indefinite integral. By Theorem 8.4 again, an indefinite integral equivalent to this second term is

$$-e^x \cos x + \int e^x \cos x\, dx,$$

and therefore an indefinite integral equivalent to $g(x)$ is

$$\bar{\bar{g}}(x) = e^x \sin x + e^x \cos x - \int e^x \cos x\, dx.$$

Thus, the difference $g(x) - \bar{\bar{g}}(x)$ is constant, which we may take to be 0, and so, solving for $\int e^x \cos x\, dx$,

$$\int e^x \cos x\, dx = \frac{1}{2} e^x (\sin x + \cos x)$$

is an indefinite integral sufficient to evaluate the given integral; that is,

$$\int_0^{\pi/2} e^x \cos x\, dx = \frac{1}{2} e^{\pi/2} - \frac{1}{2} e^0 = \frac{1}{2} (e^{\pi/2} - 1).$$

EXAMPLE 7 Consider the indefinite integral $\int x^n e^x dx$, $n \geq 1$. Identify

$$u(x) = x^n, \quad Dv(x) = e^x.$$

Then

$$Du(x) = nx^{n-1}, \quad v(x) = e^x,$$

(with $k = 0$) and

$$\int x^n e^x dx = x^n e^x - n \int x^{n-1} e^x dx.$$

Clearly, this integration by parts can be reapplied to the new indefinite integral, and in a finite number of such applications, the power function x^n will be reduced to a constant function.

8.3 PROBLEMS

1. Find an indefinite integral for each of the following.

   (a) $\int x^3 \log x\, dx$ (b) $\int x \sin 2x\, dx$

   (c) $\int x^2 \cos 3x\, dx$ (d) $\int \arcsin (4/x)\, dx$

   (e) $\int xe^{5x}\, dx$ (f) $\int e^{3x} \cos 4x\, dx$

2. Evaluate each, using integration by parts.

   (a) $\int_{\pi/2}^{0} x \cos x\, dx$ (b) $\int_0^{\log 2} xe^x\, dx$

   (c) $\int_0^{-1} x\sqrt{1+x}\, dx$ (d) $\int_0^1 \arcsin x\, dx$

   (e) $\int_{\pi/4}^{\pi/2} \sin^2 x\, dx$ (f) $\int_{\pi/6}^{\pi/3} x^2 \sin x\, dx$

3. Show that if $n \neq 1$, integration by parts can be used to prove

$$\int \sec^n x\, dx = \frac{1}{n-1}\left[\sec^{n-2} x \tan x + (n-2)\int \sec^{n-2} x\, dx\right].$$

Use this formula to find

$$\int_0^{\pi/4} \sec^6 x\, dx.$$

4. As in Problem 3, prove that if $m + n \neq 0$,

$$\int \sin^m x \cos^n x\, dx = \frac{1}{m+n}\left[\sin^{m+1} x \cos^{n-1} x + (n-1)\int \sin^m x \cos^{n-2} x\, dx\right].$$

Use this formula to find

$$\int_0^{\pi/2} \sin^3 x \cos^4 x\, dx.$$

5. As in Problem 3, prove that if $m + np + 1 \neq 0$,

$$\int x^m (ax^n + b)^p\, dx = \frac{1}{a(m+np+1)}\left[x^{m-n+1}(ax^n+b)^{p+1} - (m-n+1)b\int x^{m-n}(ax^n+b)^p\, dx\right].$$

Use this formula to find

$$\int_0^1 x^{13}(2x^7+3)^5\, dx.$$

8.4 PARTIAL FRACTION DECOMPOSITION OF A RATIONAL FUNCTION

Consider the integral $\int_a^b R$, in which R is a rational function on the interval $[a,b]$. Then there is an indefinite integral $\int R(x)dx$ which is expressible in elementary functions. We will not prove this fact, but you can easily find a proof in the mathematical literature,[6] and when you do, you will find it to be a rather tedious, systematic, examination of many possible cases. However, it is useful to know how such an indefinite integral can be found, and a few examples will convey the idea.

For such a function R there exist polynomials P and Q so that

$$R(x) = \frac{P(x)}{Q(x)} \qquad Q(x) \neq 0, \qquad a \leqslant x \leqslant b.$$

[6] For example, William F. Osgood, *Advanced Calculus,* Chapter 1, The Macmillan Company, New York, 1925.

The partial fraction theorem from algebra says that $R(x)$ has a formula which is a sum of canonical rational fractions, each of which is one of

(a) a polynomial

(b) $\dfrac{A}{(x+c)^p}$, p a positive integer, c a real number

(c) $\dfrac{Ax+B}{(x^2+bx+c)^q}$, q a positive integer, b,c real numbers.

We display a few examples to illustrate how the possible cases which can arise may be treated, and we leave the completion of the project to you.

EXAMPLE 1 $\displaystyle\int R(x)dx = \int \frac{x^4+2}{x^3+2x^2-2x+3}\,dx$

Decompose the integrand as follows. Since the degree of the numerator polynomial is larger than that of the denominator polynomial, we perform a division to get

$$R(x) = x - 2 + \frac{6x^2-7x+8}{x^3+2x^2-2x+3},$$

in which the fraction term has a numerator with lower degree than the denominator. The denominator factors

$$x^3+2x^2-2x+3 = (x+3)(x^2-x+1)$$

and the partial fraction theorem asserts that

$$\frac{6x^2-7x+8}{x^3+2x^2-2x+3} = \frac{A}{x+3} + \frac{Bx+C}{x^2-x+1}$$

for some numbers A, B, and C. If these *partial fractions* were added together, the numerator of the sum would satisfy

$$6x^2-7x+8 = (x^2-x+1)A + (x+3)(Bx+C)$$

for more than two values of x. Therefore, the coefficients of like powers of x must be the same on both sides of this equation.

$$\begin{aligned} A + B \qquad\quad &= 6 \\ -A + 3B + C &= -7 \\ A + 3C \qquad &= 8 \end{aligned}$$

Solving this system,

$$A = \frac{83}{13}, \qquad B = -\frac{5}{13}, \qquad C = \frac{7}{13}$$

and an indefinite integral is

$$\int R(x)\,dx = \int (x-2)\,dx + \frac{83}{13}\int \frac{1}{x+3}\,dx - \frac{1}{13}\int \frac{5x-7}{x^2-x+1}\,dx.$$

Of these three indefinite integrals only the last requires further treatment. If we complete the square in the denominator of the last integrand, we see that

$$\frac{5x-7}{x^2-x+1} = \frac{5x-7}{(x-\frac{1}{2})^2+\frac{3}{4}} = \frac{5}{2}\left[\frac{2(x-\frac{1}{2})}{(x-\frac{1}{2})^2+\frac{3}{4}} - \frac{9}{5}\,\frac{1}{(x-\frac{1}{2})^2+\frac{3}{4}}\right],$$

where the fraction has been broken apart (in the brackets), so that the first term is recognizable as an exact differential. There remains only the second bracketed fraction, whose indefinite integral is

$$\int \frac{1}{(x-\frac{1}{2})^2+\frac{3}{4}}\,dx.$$

For this one, the substitution (in variable notation)

$$\frac{\sqrt{3}}{2}\,t = x - \frac{1}{2}, \qquad \text{so } dx = \frac{\sqrt{3}}{2}\,dt$$

suggests itself, and

$$\begin{aligned}\int \frac{1}{(x-\frac{1}{2})^2+\frac{3}{4}}\,dx = \frac{2}{\sqrt{3}}\int \frac{1}{t^2+1}\,dt &= \frac{2}{\sqrt{3}}\arctan t \\ &= \frac{2}{\sqrt{3}}\arctan \frac{2}{\sqrt{3}}\left(x-\frac{1}{2}\right)\end{aligned}$$

is an indefinite integral for the second bracketed fraction. Hence, putting these together,

$$\begin{aligned}\int R(x)\,dx = \left(\frac{1}{2}x^2 - 2x\right) + \frac{83}{13}\log|x+3| - \frac{5}{26}\log\left|\left(x-\frac{1}{2}\right)^2+\frac{3}{4}\right| \\ - \frac{3\sqrt{3}}{13}\arctan\frac{2}{\sqrt{3}}\left(x-\frac{1}{2}\right).\end{aligned}$$

The denominator of $R(x)$ is 0 only at $x=-3$; therefore, this indefinite integral is available to compute any $\int_a^b R$ as long as the interval $[a,b]$ does not contain -3.

EXAMPLE 2 $\displaystyle\int R(x)\,dx = \int \frac{2x+1}{x^2-2x+1}\,dx$

In this case the denominator of $R(x)$ is $(x-1)^2$, so that $R(x)$ could be the sum of

$$R(x) = \frac{A}{x-1} + \frac{B}{(x-1)^2}.$$

Hence,

$$2x + 1 = A(x - 1) + B$$

and clearly, $A = 2$, $B = 3$. Hence,

$$\int R(x)dx = 2\int \frac{1}{x-1}\,dx + 3\int \frac{1}{(x-1)^2}\,dx = 2\log|x-1| - \frac{3}{x-1}$$

is an indefinite integral for $\int_a^b R$ over any interval $[a,b]$ which does not contain 1.

When the denominator polynomial contains a factor $(x+c)^p$, $p > 1$, the partial fraction decomposition must be assumed to contain all of the fractions $\frac{A_j}{(x+c)^j}$, $j = 1,2, \ldots, p$. The same remark should be made for quadratic factors.

EXAMPLE 3 $\int R(x)dx = \int \frac{x^2}{(1+x^2)^2}\,dx$

The trial partial fractions are

$$\frac{x^2}{(1+x^2)^2} = \frac{Ax+B}{1+x^2} + \frac{Cx+D}{(1+x^2)^2},$$

where $A = 0$, $B = 1$, $C = 0$, and $D = -1$; and an indefinite integral is

$$\int R(x)dx = \int \frac{1}{1+x^2}\,dx - \int \frac{1}{(1+x^2)^2}\,dx.$$

For the second term substitute $x = \tan t$. Then compute an indefinite integral for this second term, with a free use of variables, to get

$$\begin{aligned}\int \frac{1}{(1+x^2)^2}\,dx &= \int \cos^2 t\,dt = \frac{1}{2}\int (1+\cos 2t)\,dt \\ &= \frac{1}{2}t + \frac{1}{2}\sin t \cos t \\ &= \frac{1}{2}\arctan x + \frac{1}{2}\frac{x}{1+x^2}.\end{aligned}$$

Thus, an indefinite integral for this problem is

$$\int R(x)dx = \frac{1}{2}\arctan x - \frac{1}{2}\frac{x}{1+x^2}.$$

8.4 PROBLEMS

1. Express each of the following rational functions in partial fractions.

(a) $\frac{x-2}{(x+1)(x-3)}$ (b) $\frac{x^2-2}{(x+1)(x-3)}$

(c) $\dfrac{1}{(x+1)(x-3)}$ (d) $\dfrac{x-2}{(x+1)^2\,(x-3)}$

(e) $\dfrac{x-2}{(x+1)^2\,(x-3)^3}$

2. Proceed as in Problem 1 for the following.

(a) $\dfrac{x+4}{(x+2)(x^2-2x+3)}$ (b) $\dfrac{x^3+4}{(x+2)(x^2-2x+3)}$

(c) $\dfrac{x+4}{(x+2)(x^2-2x+3)^2}$ (d) $\dfrac{x+4}{(x+2)^3\,(x^2-2x+3)^2}$

3. Find an indefinite integral for each rational function in 1 and 2.

8.5 TABLES OF INTEGRALS

Our discussion of the technique of integration has only introduced the subject, and we have used it to gain some acquaintance with ways to solve $Dg=f$ for g, given f. Usually, if a solution is not readily apparent, consult a table of integrals,[7] in which are found indefinite integrals for each of a large class of functions. When you are confronted with an integral which is not recognizable, it usually pays to try a substitution, or apply integration by parts, and keep a table of integrals at hand in which the by-products may turn up. There is an especially good summary of the methods that are usually effective in F. L. Griffin *Mathematical Analysis, Higher Course,* Houghton Mifflin, New York, 1927, Chapter IV. This is written in variable notation, and although the chapter is brief, it is remarkably complete. If this technique attracts you, the problems in that chapter will be enjoyable.

8.5 PROBLEMS

1. Derive each of the following formulas. (These can be checked easily by differentiation, but you should also be able to perform the integrations.)

(a) $\displaystyle\int a^x\,dx=\frac{a^x}{\log a},\quad a>0$

(b) $\displaystyle\int (1/x)\,dx=\log|x|,\quad x\neq 0$

(c) $\displaystyle\int \tan x\,dx=\log|\sec x|$

(d) $\displaystyle\int \cot x\,dx=\log|\sin x|$

(e) $\displaystyle\int \sec x\,dx=\log|\sec x+\tan x|$

[7] For example, H. B. Dwight, *Tables of Integrals and Other Mathematical Data,* The Macmillan Company, New York, 1934.

(f) $\int \csc x \, dx = \log |\csc x - \cot x|$

(g) $\int \frac{dx}{\sqrt{a^2 - x^2}} = \arcsin \frac{x}{a}, \quad x^2 < a^2$

(h) $\int \frac{dx}{a^2 + x^2} = \frac{1}{a} \arctan \frac{x}{a}$

(i) $\int \frac{dx}{x\sqrt{x^2 - a^2}} = \frac{1}{a} \operatorname{arcsec} \frac{x}{a}, \quad x^2 > a^2$

(j) $\int \frac{dx}{u^2 - a^2} = \frac{1}{2a} \log \left|\frac{u - a}{u + a}\right|$

(k) $\int \frac{dx}{\sqrt{x^2 \pm a^2}} = \log |x + \sqrt{x^2 \pm a^2}|, \quad x^2 > a^2$ in the $x^2 - a^2$ case

(l) $\int \sqrt{a^2 - x^2} \, dx = \frac{1}{2}[x\sqrt{a^2 - x^2} + a^2 \arcsin(x/a)]$

(m) $\int \sqrt{x^2 \pm a^2} \, dx = \frac{1}{2}[x\sqrt{x^2 \pm a^2} + a^2 \log |x + \sqrt{x^2 \pm a^2}|]$

REVIEW PROBLEMS

1. Prove that if g is a differentiable function and f is a continuous function, then

$$Dg = f, \quad g(a) = k$$

if, and only if,

$$g(x) = \int_a^x f + k.$$

2. Prove that if a function g has the rule

(A) $$g(x) = e^{cx}\left[\int_a^x e^{-ct} f(t) \, dt + ke^{-ca}\right], \quad a \leqslant x \leqslant b$$

then

(B) $$Dg - cg = f, \quad g(a) = k.$$

3. Prove that if a differentiable function $\bar{g}$ satisfies (B) of Problem 2 and if f is continuous, then $\bar{g} = g$ defined in (A) of Problem 2.
4. Using the results of the preceding problems, solve each of the following *differential equations* for g.
   (a) $Dg(x) = e^x, \quad g(0) = 1$
   (b) $Dg(x) = e^x, \quad g(0) = 5$
   (c) $Dg(x) = e^x, \quad g(1) = 3$

(d) $Dg(x) - 2g(x) = e^x, \quad g(0) = -1$
(e) $Dg(x) - 2g(x) = e^x, \quad g(1) = 0$
(f) $Dg(x) - g(x) = \cos x, \quad g(\pi/2) = \frac{1}{2}$
(g) $Dg(x) + 4g(x) = 3x, \quad g(0) = 0$

5. In Problem 2 if $k = 0$ in line (B), then show that the solution of the differential equation (B) is

$$\text{(A}'\text{)} \qquad g(x) = \int_a^x e^{c(x-t)} f(t)\, dt.$$

Solve each of the following differential equations.
(a) $Dg(x) - 2g(x) = 3, \quad g(1) = 0$
(b) $Dg(x) - 2g(x) = 3x, \quad g(1) = 0$
(c) $Dg(x) - 2g(x) = e^x, \quad g(0) = 0$
(d) $Dg(x) - 2g(x) = e^{2x}, \quad g(0) = 0$
(e) $Dg(x) - 2g(x) = \sin x, \quad g(\pi/2) = 0$

6. The device described in Problem 2 for the solution of a differential equation can be extended to apply to differential equations which contain derivatives of higher order. We illustrate this extension for the *second order* differential equation, in which the second derivative of the function g occurs.

Prove that if a function g has the rule

$$\text{(C)} \quad g(x) = e^{qx} \int_a^x e^{-qu} \left[e^{pu} \int_a^u e^{-pt} f(t)\, dt \right] du, \quad a \leqslant x \leqslant b,$$

then[8]

$$\text{(D)} \qquad D^2 g - (p + q) Dg + pqg = f, \quad g(a) = Dg(a) = 0.$$

7. Prove that if a twice differentiable function $\bar{g}$ satisfies (D) of Problem 6 and if f is continuous, then $\bar{g} = g$ defined in (C).

8. Using the results of Problems 6 and 7, solve each of the second order differential equations for g.
(a) $D^2 g(x) + Dg(x) - 2g(x) = x, \quad g(1) = Dg(1) = 0$
(Notice that the associated polynomial $y^2 + y - 2 = (y - 1)(y + 2)$, so in this example $p = 1$, $q = -2$. In the following problems this associated polynomial will always have real factors which give p and q. The method is, in fact, more general than this.)
(b) $D^2 g(x) + Dg(x) - 2g(x) = e^{-x}, \quad g(0) = Dg(0) = 0$
(c) $D^2 g(x) + Dg(x) - 2g(x) = \cos x, \quad g(\pi/2) = Dg(\pi/2) = 0$
(d) $D^2 g(x) + Dg(x) - 2g(x) = x + e^{-x}, \quad g(0) = Dg(0) = 0$

9. In Problem 6 show that the solution of the differential equation (D) is

[8] The letters f and g stand for functions, while p and q stand for real numbers.

(E) $$g(x) = e^{qx} \int_a^x \left[e^{(p-q)u} \int_a^u e^{-pt} f(t)\, dt \right] du.$$

10. Prove that the function g of line (E) in Problem 9 has the formula (using a single integral)

(F) $$\begin{aligned} g(x) &= \frac{1}{p-q} \int_a^x [e^{p(x-t)} - e^{q(x-t)}] f(t)\, dt, \quad p \neq q \\ &= \int_a^x (x-t) e^{q(x-t)} f(t)\, dt, \quad p = q. \end{aligned}$$

Hint: Integrate the outside integral of line (E) by parts.

Appendix I
SETS

We often observe different objects which share a certain attribute, or have the same property in common. Thus, if we meet someone and discover that he believes the world is round, he shares this belief with us. If another person thinks the world is not round he is, in this regard at least, different from ourselves. In set language we say that the one person belongs to the *set* of people who think the world is round, whereas the other person does not belong to that set.

In most real-life situations a use of formal set language is not necessary. We can speak of all enlightened people, or all unenlightened people —and these may be all Republicans or all non-Republicans, according to our persuasion. But in mathematics we frequently must define rather complicated sets, and it is necessary to construct a more refined technique to deal with them.

We agree that the term "set" will be applied only to cases for which any object under consideration either possesses the property in which we are interested or does not possess it. In this light probably all citizens of the United States who are Republicans would not form an admissible set. There are too many shades of opinion about politics to enable us to classify every citizen as a Republican or not. But if we consider all registered voters in the United States who are Republicans, then the registration rolls can be used to define this collection as a set. Particular sets which are used in this book are sets of mathematical objects, such as numbers or functions, and properties assigned to these objects are carefully specified by definitions.

If we assume that the term "collection of objects" has a meaning, then we define a *set* to be a collection of objects for which it can be determined whether an object is in the collection or is not in the collection. For example, the collection A of positive whole numbers not exceeding 9 is the set

$$A = \{1,2,3,4,5,6,7,8,9\}.$$

For small sets, symbols for the members of the collection, enclosed in braces, form a symbol for the set. Thus,

$$B = \{2,4,6,8,10,12\}$$

is the set of positive even whole numbers not exceeding 13.

If a set is large and has many members, this kind of symbol is unwieldy, and we employ the *set-builder* notation. This is self-explanatory when we write the preceding examples as

$$A = \{x | 1 \leq x \leq 9, \quad x \text{ is a whole number}\}$$
$$B = \{x | x = 2p,\ 1 \leq p \leq 6, \quad p \text{ is a whole number}\}.$$

A member of the collection is called an *element* of the set, and an element of a set *belongs* to it. Sometimes a listing of the elements is clearer than the set-builder formula (as in B above), but when the set is large the set-builder formula usually is preferable.

NOTATION

In order to write about sets the following notation has become standard. Let S and T be sets; then

$$x \in S \text{ means } x \text{ is an element of } S.$$

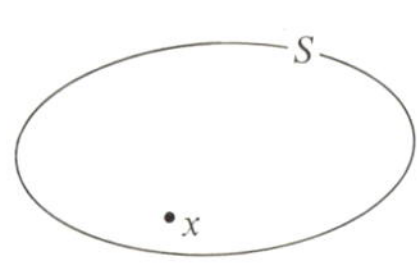

$$x \notin S \text{ means } x \text{ is not an element of } S.$$

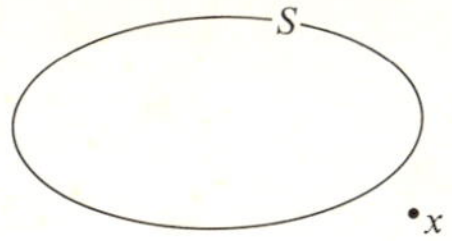

$S \subset T$ means that if $x \in S$, then $x \in T$ also.

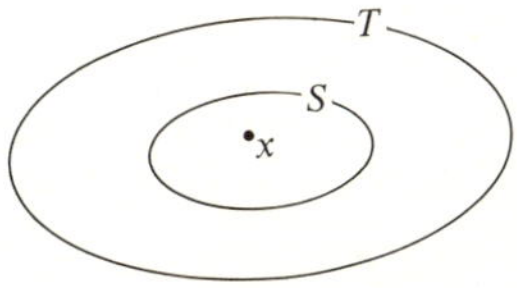

The set S in this case is called a *subset* of the set T. Any element of S is an element of T. If both $S \subset T$ and $T \subset S$ are true, then clearly S and T are alternative symbols for the same set, and we write this fact as $S = T$.

$$S \not\subset T \text{ means that } S \subset T \text{ is false.}$$

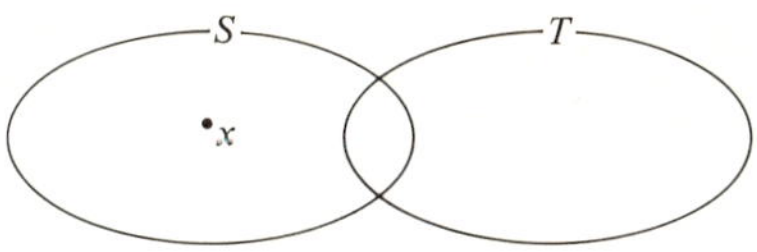

That is, there is some x such that $x \in S$ and $x \notin T$.

$$S \cap T = \{x | x \in S \quad \text{and} \quad x \in T\}$$

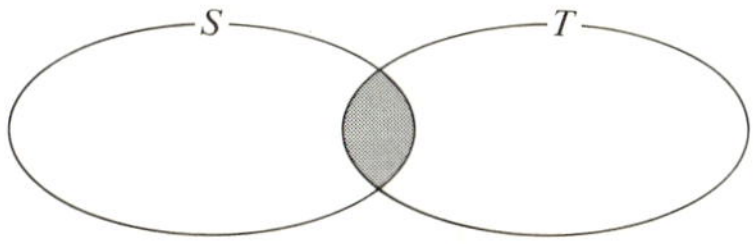

$S \cap T$

is the *intersection* of S and T. It contains just those elements common to S and T.

$$S \cup T = \{x | x \in S \quad \text{or} \quad x \in T, \quad \text{or both}\}$$

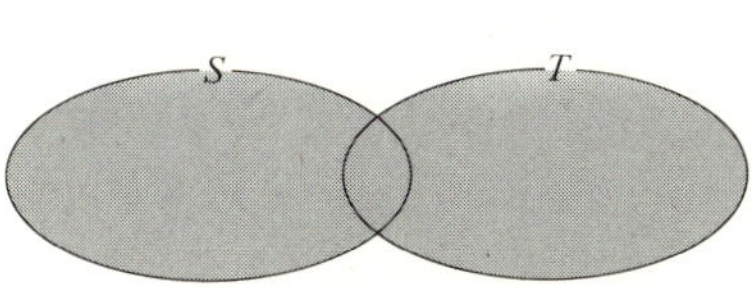

$S \cup T$

is the *union* of S and T. It contains all the elements of S and T taken together.

$$S - T = \{x | x \in S \quad \text{and} \quad x \notin T\}$$

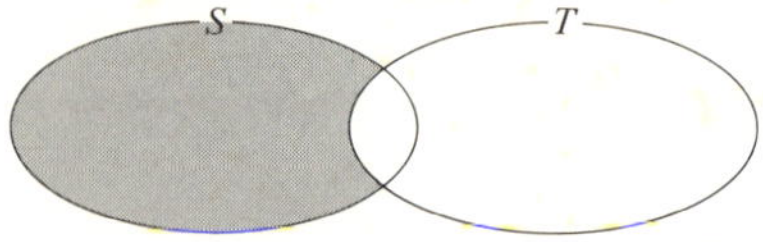

is the *complement* of T in S. It contains those elements of S which are not in T. In certain discussions it is understood that only a certain kind of object is being considered. For example, in plane geometry attention is devoted to the set of points in a particular plane. We call this set the immediate *universe of discourse U*, and nothing else exists for us at the moment. Then if T is a set of points in that plane, the set

$$cT = U - T$$

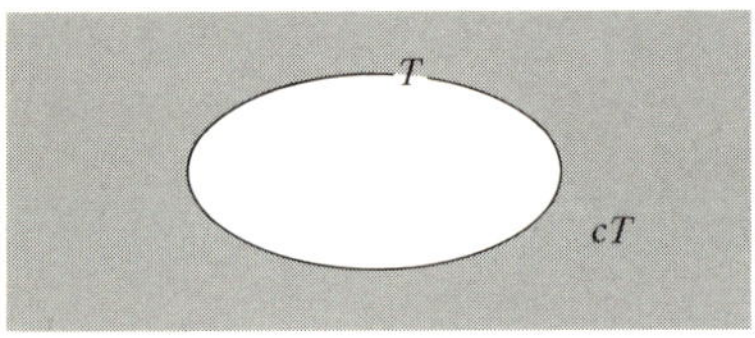

is referred to as the *complement* of T. In the universe of real numbers if T is the set of rational numbers, then cT is the set of irrational numbers.

THE EMPTY SET

Consider the set $\varnothing$,

$$\varnothing = \{x | x = 0 \quad \text{and} \quad x = 1, \quad x \text{ is a whole number}\},$$

and choose any object s. If s is not a whole number, then $s \notin \varnothing$. Otherwise s is a whole number. If s is not 0, then $s \notin \varnothing$. Finally, if $s = 0$, then $s \neq 1$ and $s \notin \varnothing$. Thus, for any object s (and we have covered them all) we have determined whether or not $s \in \varnothing$ (it happens that $s \notin \varnothing$ is always true), and so $\varnothing$ is a set. It is called the *empty* set or the *null* set. You can show that for any set S, $\varnothing \subset S$ is true. If we define

$$\varnothing = \{x | x \text{ has contradictory properties}\},$$

then we get the same empty set as before; that is, there is only one empty set. Also you can verify readily that for any set S,

$$\varnothing \subset S$$
$$\varnothing \cap S = \varnothing, \quad \varnothing \cup S = S$$
$$S - \varnothing = S, \quad \varnothing - S = \varnothing$$

and in any universe of discourse U,

$$c\varnothing = U, \qquad cU = \varnothing.$$

TECHNIQUE

Useful set identities can be constructed from the definitions in the notation section. Their truth is usually clear from the definitions and their accompanying diagrams. To illustrate this, consider some universe U and the sets S and T, each a subset of U. Then

$$\text{(A)} \qquad c(S \cup T) = cS \cap cT.$$

From Figure I.1 we see that the left side of the Identity (A) is the set of all elements in the rectangle outside of the linked oval patches. The right side of the Identity (A) is the set of all elements of U both outside of the S oval and outside of the T oval—clearly the same set. Thus, we are convinced that there ought to be a formal proof of the Identity (A) based on the definitions of union, intersection, and complement. Very likely you practiced this already with identities above which involved the empty set.

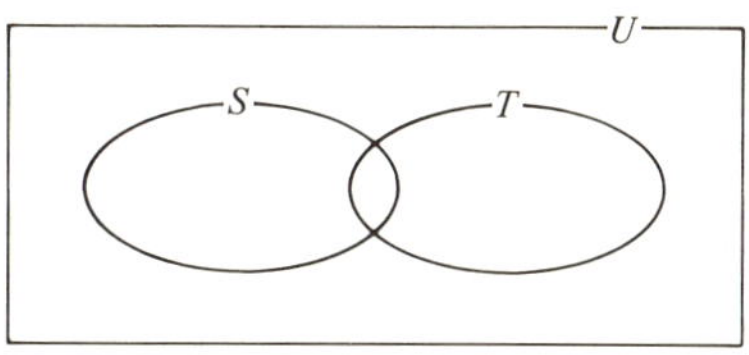

Figure I.1

We give a proof of the Identity (A).

Proof: Write (A) as

$$L = U - (S \cup T) \qquad \text{and} \qquad (U - S) \cap (U - T) = R.$$

If $x \in L$, then $x \in U$ and $x \notin S \cup T$. That is, $x \in U$, $x \notin S$, and $x \notin T$. This is the same as saying that $x \in U$ and $x \notin S$, so that $x \in (U - S)$, and also that $x \in U$ and $x \notin T$, so that $x \in (U - T)$. Therefore, $x \in (U - S) \cap (U - T)$ and $x \in R$. We have proved that $L \subset R$.

Now suppose that $x \in R$. Then both $x \in (U - S)$ and $x \in (U - T)$ are true, so that $x \in U$, $x \notin S$, and $x \notin T$; in particular, $x \notin (S \cup T)$. Then $x \in U - (S \cup T)$ so that $x \in L$. We have proved that $R \subset L$.

Since both $L \subset R$ and $R \subset L$ are true, then $L = R$. ■

The most common application of sets to analysis involves a demonstration that two definitions of particular sets are, in fact, definitions of the same set. To do this we give a different name to each set produced by a definition, and then show that each is a subset of the other. This, in turn, is done by considering an arbitrary element of one set and chasing it around until it turns up in the other set to establish the desired inclusion relation. We used this device in the proof above.

COMMENTS

It is especially important to appreciate the distinction between an element of a set and the set itself. The chocolates in a box are not the box of chocolates. Suppose that we assume a metalanguage of sets, and the existence of the empty set $\varnothing$. The following exercise illustrates the point we are making.

Although the empty set $\varnothing$ contains no elements, the set whose single element is $\varnothing$, namely $\{\varnothing\}$, is nonempty. Using a familiar numeral as a name, we write

$$1 = \{\varnothing\}.$$

Since $\varnothing$ and $\{\varnothing\}$ are distinct objects, the set

$$2 = \{\varnothing, \{\varnothing\}\}$$

is a new set. Similarly, the set

$$3 = \{\varnothing, \{\varnothing\}, \{\varnothing, \{\varnothing\}\}\}$$

is still another. If n denotes a set, define

$$n + 1 = n \cup \{n\}.$$

We call a set S an *I-set* if

(a) $1(=\{\varnothing\}) \in S$
(b) $n \in S$ implies $n + 1(= n \cup \{n\}) \in S$.

Thus an I-set is inductive as described in Chapter 1, Review Problem 18. Therefore, as in Chapter 1, Review Problem 19, the intersection of all I-sets is essentially the set of positive whole numbers. Starting with the empty set $\varnothing$, the positive whole number system can be contructed in this way, and in fact, the real and complex number systems can then be built from it.

You surely have noticed that the set of positive whole numbers can be put in one-to-one correspondence with the set of even positive whole numbers (the odd whole numbers have been discarded).

$$n \longleftrightarrow 2n, \quad n = 1,2,3, \ldots .$$

A proper subset of such an I-set may have as "many" elements as does the whole I-set. Many other sets have this property, too. They are called *infinite* sets. A set which is not infinite is *finite*. The set R of real numbers is infinite, as also is any open interval of real numbers.

Appendix II NOTATION FOR SUMS

If one were directed to sum the odd numbers from 3 through 13, he could write

$$3 + 5 + 7 + 9 + 11 + 13 = 48.$$

Both the left- and the right-hand members of this equation are symbols for the number 48. The left-hand member, though longer, does record how the number 48 was reached—in this instance by adding up a certain set of odd numbers.

Another way to describe this operation could be, "Add together all numbers of the form $2j + 1$ for $j = 1, 2, 3, 4, 5$, and 6." This is what the symbol

$$\sum_{j=1}^{6} (2j + 1)$$

means. More generally, if the number a_j is defined for each positive whole number j in an interval $m \leqslant j \leqslant n$, then the symbol

$$\sum_{j=m}^{n} a_j$$

stands for the sum of all these defined a_j's; that is,

$$a_m + a_{m+1} + a_{m+2} + \ . \ . \ . \ + a_n.$$

Formally, we define *summation* as follows. Let a function whose domain is the set of positive whole numbers have the rule

$$j \longrightarrow a_j.$$

Then

$$\sum_{j=1}^{1} a_j = a_1$$
$$\sum_{j=1}^{n} a_j = \sum_{j=1}^{n-1} a_j + a_n, \quad 1 < n$$
$$\sum_{j=m}^{n} a_j = \sum_{j=1}^{n} a_j - \sum_{j=1}^{m-1} a_j, \quad 2 \leqslant m \leqslant n.$$

This is an inductive definition; it defines the symbol $\sum_{j=1}^{n} a_j$ for $n = 1$, and for larger n gives the value of the symbol in terms of its value for $n - 1$. The third statement extends the symbol for certain useful applications.

There are three essential properties of the summation symbol.

(a) $\sum_{j=1}^{n} ca_j = c \sum_{j=1}^{n} a_j, \quad 1 \leqslant n$

(b) $\sum_{j=1}^{n} (a_j + b_j) = \sum_{j=1}^{n} a_j + \sum_{j=1}^{n} b_j, \quad 1 \leqslant n$

(c) $\sum_{j=1}^{n} a_j = \sum_{j=p+1}^{n+p} a_{j-p}, \quad 0 \leqslant p, \quad 1 \leqslant n.$

These properties are clear when the summation symbol is viewed as an abbreviation for the addition of a string of terms. They can be derived from the formal definitions by mathematical induction. We exhibit this derivation for properties (a) and (c), leaving (b) to you.

Proof: (a) If $n = 1$, then

$$\sum_{j=1}^{n} ca_j = \sum_{j=1}^{1} ca_j = ca_1 = c \sum_{j=1}^{1} a_j = c \sum_{j=1}^{n} a_j.$$

If $n > 1$, then, assuming (a) for $n - 1$,

$$\sum_{j=1}^{n} ca_j = \sum_{j=1}^{n-1} ca_j + ca_n = c\sum_{j=1}^{n-1} a_j + ca_n$$
$$= c\left[\sum_{j=1}^{n-1} a_j + a_n\right] = c\sum_{j=1}^{n} a_j. \blacksquare$$

Proof: (c) Notice that if $m \geqslant 2$,

$$\sum_{j=m}^{m} a_j = \sum_{j=1}^{m} a_j - \sum_{j=1}^{m-1} a_j = \left[\sum_{j=1}^{m-1} a_j + a_m\right] - \sum_{j=1}^{m-1} a_j = a_m.$$

Now, if $n = 1$,

$$\sum_{j=1}^{n} a_j = \sum_{j=1}^{1} a_j = a_1 = a_{(p+1)-p} = \sum_{j=p+1}^{1+p} a_{j-p} = \sum_{j=p+1}^{n+p} a_{j-p}.$$

Notice, furthermore, that if $2 \leqslant m \leqslant n$, then

$$\sum_{j=m}^{n} a_j = \sum_{j=1}^{n} a_j - \sum_{j=1}^{m-1} a_j = \sum_{j=1}^{n-1} a_j + a_n - \sum_{j=1}^{m-1} a_j = \sum_{j=m}^{n-1} a_j + a_n.$$

Therefore, if $n > 1$, assuming (c) for $n - 1$,

$$\sum_{j=1}^{n} a_j = \sum_{j=1}^{n-1} a_j + a_n = \sum_{j=p+1}^{n-1+p} a_{j-p} + a_{(n+p)-p} = \sum_{j=p+1}^{n+p} a_{j-p}. \blacksquare$$

In the last line of the proof of (c) above you see that the subscript on the term a_{j-p} is $j - p$ and that under Σ we find $j - p = 1$. We emphasize that this index is a dummy; that is, any letter may be used for it. The two symbols

$$\sum_{j=1}^{n} a_j \qquad \text{and} \qquad \sum_{k=1}^{n} a_k$$

stand for the same number. This should be clear enough, but it, too, can be verified.

Proof: If $n = 1$, then

$$\sum_{j=1}^{n} a_j = \sum_{j=1}^{1} a_j = a_1 = \sum_{k=1}^{1} a_k = \sum_{k=1}^{n} a_k.$$

If $n > 1$, assuming the truth of the case $n - 1$, we find

$$\sum_{j=1}^{n} a_j = \sum_{j=1}^{n-1} a_j + a_n = \sum_{k=1}^{n-1} a_k + a_n = \sum_{k=1}^{n} a_k. \blacksquare$$

Other useful symbols can be treated as we have described the Σ symbol here. For example, define $\prod_{j=1}^{n} a_j$, the product symbol, by

$$\prod_{j=1}^{1} a_j = a_1$$

$$\prod_{j=1}^{n} a_j = a_n \prod_{j=1}^{n-1} a_j.$$

This represents the product of n factors multiplied together.

If the symbols a_j stand for sets, then

$$\bigcup_{j=1}^{n} a_j \quad \text{and} \quad \bigcap_{j=1}^{n} a_j$$

have analogous definitions which you can construct and study.

Appendix III VARIABLES

VARIABLES

The word "variable" is used conventionally as a synonym for the word "symbol" when that symbol refers to an unspecified element of some set. In our discussion of velocity in Chapter 3 we employ the formula $s = 16t^2$, in which t stands for any number in a set of measurements of time, and s stands for any number in a set of position measurements. In a quite imprecise way these letters are said to "vary" over the number sets, and they are called *variables*.

Although this language is unnecessary for formal mathematical work, it is used in applications of mathematics especially. It is used by, and is useful for, most working mathematicians, so we should look for a way to give it a precise meaning. We now illustrate one way to do this by constructing an interpretation of the formula $s = 16t^2$ as it describes the behavior of a falling body. We call a function a *variable* when it occurs in a context such as the following.

Let us postulate, in the physical world, the existence of a set of objects called "times" or "instants," and another set called "positions."

$$\{\text{times } T\}, \qquad \{\text{positions } P\}$$

Suppose that we have at hand two instruments: a clock which assigns to each time T a number $t(T)$, and a ruler which assigns to each position P a number $s(P)$. The letter t, then, denotes a function whose domain is $\{\text{times } T\}$ and whose range is a set of real numbers. The letter s likewise denotes a function with domain $\{\text{positions } P\}$ and with a set of real numbers for its range. These two functions, t and s, are variables in this discussion. We assume that the variables, and other functions we use, have inverses. In most applications the domain of any variable can be chosen suitably, so this assumption is not a crippling limitation.

We perform an experiment in which an object falls from rest, and it is observed by both of the instruments. They are set to record their numerical readings simultaneously, so the print-out is a tabulation of number pairs $t(T)$, $s(P)$. We observe that this set of number pairs is the graph of a function f whose formula is

$$f(x) = 16x^2,$$

that is,

$$\text{(A)} \qquad s(P) = 16(t(T))^2.$$

Equation (A) holds for each entry in the print-out. Each entry, furthermore, picks out a particular time T and a particular position P from the original sets of physical objects and matches them. Call this correspondence ϕ, a function whose domain is $\{\text{times } T\}$ and whose range is $\{\text{positions } P\}$,

$$P = \phi(T), \quad T \in \{\text{times } T\}.$$

Equation (A) is, then,

$$s(\phi(T)) = f(t(T))$$

or, in function notation,

$$\text{(B)} \qquad s \circ \phi = f \circ t, \quad \text{domain is } \{\text{times } T\}.$$

(See Figure III.1.)

Formula (B) expresses the fundamental relation connecting the variables s and t, which translate the physical quantities into numbers by measuring them, with the observed relation f between their measures and the induced mapping ϕ between the physical quantities themselves. The function ϕ can be identified formally by solving (B),

$$\phi = s^{-1} \circ f \circ t.$$

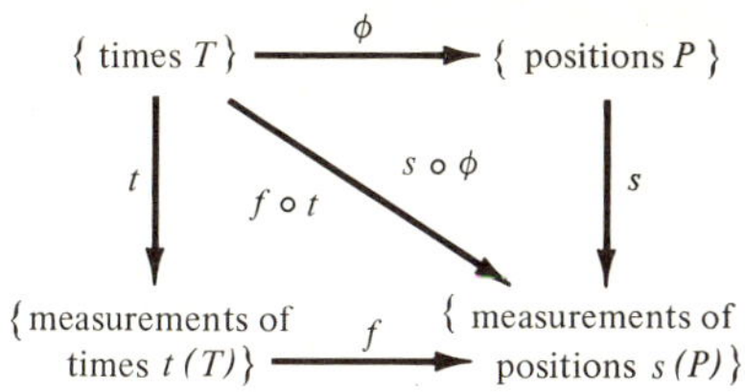

Figure III.1

Hence, if the measuring instruments, that is, the variables s and t, are given, and the law connecting the measures is known, then the relation ϕ between the physical quantities is determined. This relation ϕ has little practical importance, though, since the only way to identify physical objects quantitatively is by their measures. From a theoretical point of view, Formula (B) serves to define operationally what the relation ϕ is.

Let us repeat the same physical experiment, but with different instruments. Before, we used a second-clock, and a foot-rule. This time we will use a minute-clock and a meter-rule. Thus we now use different variables, say $\bar{t}$ and $\bar{s}$, to measure time and position, respectively, and we have the result given by Figure III.2

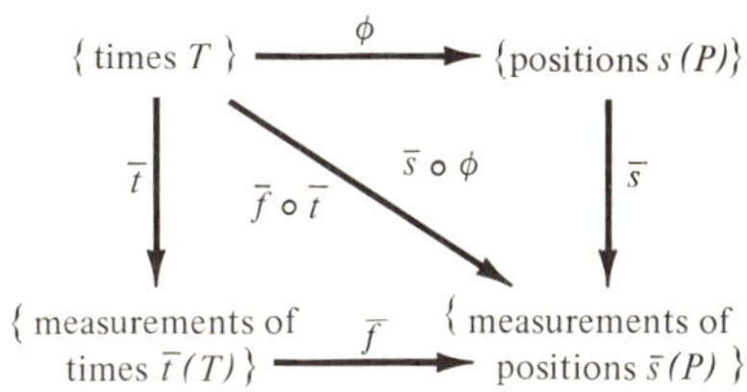

Figure III.2

where presumably $\bar{f}$ is not the same function as f. If we can deduce what $\bar{f}$ is, we will succeed in writing the "law of a falling body" in the new "units," minutes and meters.

Superimposing Figure III.2 on Figure III.1, we get Figure III.3, where we know ϕ, f, and the four variables $t, s, \bar{t}, \bar{s}$, and we seek the function $\bar{f}$. From the diagram we have

$$\bar{f} \circ \bar{t} \circ t^{-1} = \bar{s} \circ s^{-1} \circ f,$$

which can be solved for $\bar{f}$;

$$\text{(C)} \qquad \bar{f} = (\bar{s} \circ s^{-1}) \circ f \circ (t \circ \bar{t}^{-1}).$$

We note, in passing, that from (B), $s^{-1} \circ f \circ t = \phi$, so that $\bar{f} = \bar{s} \circ \phi \circ \bar{t}^{-1}$

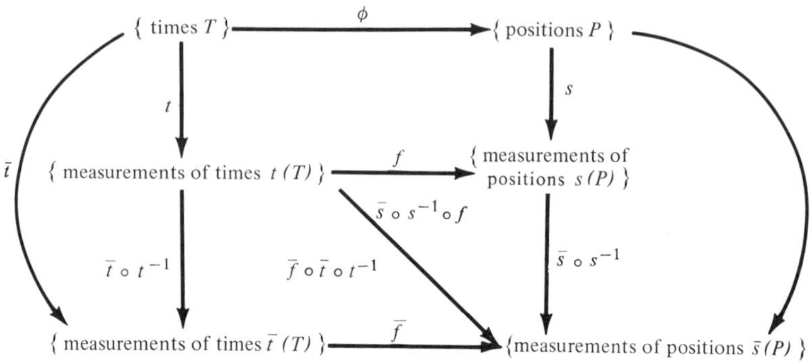

Figure III.3

and therefore

$$(\bar{\mathrm{B}}) \qquad \bar{s} \circ \phi = \bar{f} \circ \bar{t}$$

expresses the fundamental relation in the new variables for the same relation ϕ between the physical objects.

Returning to the search for $\bar{f}$, we examine the function $t \circ \bar{t}^{-1}$ which occurs in (C) and which maps the time measurements by the minute-clock to the time measurements by the second-clock. The instrument-maker tells us that

$$t \circ \bar{t}^{-1}(x) = 60x.$$

(Notice that this is the inverse of the function $\bar{t} \circ t^{-1}$ which appears on the diagram.) Similarly, for the rulers,

$$\bar{s} \circ s^{-1}(x) = 0.3048x.$$

Therefore, performing the compositions in (C),

$$\begin{aligned} \bar{f}(x) &= 0.3048[16(60x)^2] \\ &= 17556.48x^2, \end{aligned}$$

which describes the falling body as viewed by a minute-clock and a meter-stick. The physical phenomenon we are observing remains the same in both experiments, as is shown by the occurrence of the same function ϕ in both Formulas (B) and ($\bar{\mathrm{B}}$).

Surely, you have changed units in formulas arising in a science course to get new formulas, and you do it by devices which are quick and dependable. However, by treating variables as functions, we can see the application of mathematics to those physical relationships which, we presume, exist independently of the measurements we make.

In Appendix IV we meet the problem of assigning rectangular coordinates to a geometric point P. From the point of view that a geometric point P has an independent existence in the plane, its coordinates, say

(x,y), can be assigned when a pair of coordinate axes is chosen (a "reference frame"). Thus the numbers x and y depend on P, and we will interpret them as values of coordinate variables, ξ and η, functions whose common domain is the set of points in the plane. If different coordinate axes are chosen, that is, if a different reference frame is used, a different pair of coordinate variables arises, each with the same original domain of points.

Appendix IV ANALYTIC GEOMETRY

ANALYTIC PLANE

The collection of drawings in a draftsman's file illustrates one model of plane euclidean geometry. These drawings contain points, lines, and a variety of other kinds of figures. It is this model that we usually study in school geometry. Axiomatic geometry is the abstract system of which the drawings provide one model or interpretation. This begins with undefined elements, called points and lines, and contains a set of axioms from which properties of configurations of points and lines are deduced. It is immaterial to the axiomatic theory just what points and lines are.

An alternative model of plane geometry can be constructed using numbers in place of drawings as the basic elements. This model is called *analytic geometry,* although usually in practice the term analytic geometry is applied to a study of the relations between these two models. We will illustrate in this appendix how this can be done. In order to distinguish the points and lines of the analytic model from those found in drawings, we will call the latter "geometric points" and "geometric lines."

Consider the set π,

$$\pi = \{(x,y) | x,y \quad \text{real numbers}\},$$

of all ordered pairs of real numbers. Call the set π a *plane* and the elements of π, the number pairs (x,y), *points*. Certain subsets of π are

$$\Lambda(A,B,C) = \{(x,y) | Ax + By + C = 0\},$$

where the letters all stand for real numbers. If the coefficients A, B, and C are all 0, then $\Lambda(0,0,0) = \pi$. If $C \neq 0$ and $A = B = 0$, then $\Lambda(0,0,C) = \varnothing$—the set is empty. Otherwise $A^2 + B^2 \neq 0$, that is, one of A and B is not 0. If $A \neq 0$, the point $(-(C/A),0) \in \Lambda(A,B,C)$, and the point $(1 - (C/A),0) \notin \Lambda(A,B,C)$; hence this set is part, but not all, of the plane π. If $A = 0$, then $B \neq 0$, and we come to the same conclusion.

We define a *line* in the plane π to be any set $\Lambda(A,B,C)$ with $A^2 + B^2 \neq 0$. These sets can be examined by algebraic methods and can be found to have properties analogous to properties of geometric lines in a geometric plane. To begin with, notice that the two equations

$$\text{(A)} \qquad \begin{aligned} Ax + By + C &= 0 \\ \bar{A}x + \bar{B}y + \bar{C} &= 0 \end{aligned}$$

have identical solution sets $\{(x,y)\}$ whenever

$$\bar{A} = kA, \quad \bar{B} = kB, \quad \bar{C} = kC, \quad k \neq 0,$$

and so

$$\Lambda(A,B,C) = \Lambda(kA,kB,kC).$$

The two equations in (A) have no solution in common in case

$$\bar{A} = kA, \quad \bar{B} = kB, \quad \bar{C} \neq kC, \quad k \neq 0,$$

and so

(i) $\Lambda(A,B,C) \cap \Lambda(kA,kB,\bar{C}) = \varnothing$.

The two equations in (A) have exactly one solution in common in case

$$\bar{A} = kA, \quad \bar{B} = kB, \quad \text{for no } k,$$

and so

(ii) $\Lambda(A,B,C) \cap \Lambda(\bar{A},\bar{B},\bar{C}) = \{\text{single point}\}$.

Lines of case (i) are called *parallel;* lines of case (ii) are *nonparallel* and *intersect* in a single point.

The subsets of the plane π,

$$\pi_> (A,B,C) = \{(x,y)|Ax + By + C > 0\}$$
$$\pi_< (A,B,C) = \{(x,y)|Ax + By + C < 0\}$$

are called *half-planes* determined by $\Lambda(A,B,C)$, and

$$\pi_<(A,B,C) \cup \Lambda(A,B,C) \cup \pi_>(A,B,C) = \pi.$$

Continuing in this way, defining subsets of π and studying their properties, we can build a structure in these sets of ordered pairs of numbers which has a geometric character; that is, the structure can be described in familiar geometric language. Rather than proceeding in such a formal way, however, let us set up a connection between the number sets and the geometric diagrams they correspond to so that both can be studied together.

COORDINATE VARIABLES

Suppose that we use the symbol $Ax + By + C = 0$ to stand for both the algebraic equation and also for the set $\Lambda(A,B,C)$. Then we can speak of the "line $Ax + By + C = 0$" and discard the Λ symbol. When we refer to the figure on the drawing board which represents a line, we will speak of a "geometric line." Points and other geometrical entities will be denoted similarly.

Consider the two lines

$$x + 0y - p = 0, \qquad 0x + y - q = 0,$$

whose intersection is the point (p,q). See Figure IV.1. Choose two perpendicular geometric lines to correspond to these, and let the geometric

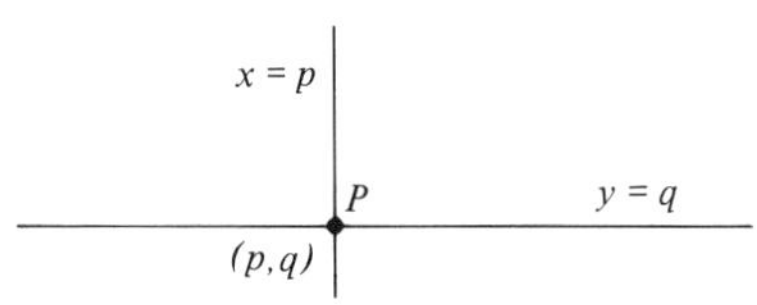

Figure IV.1

point P, their geometrical intersection, correspond to (p,q). The two lines

$$x + 0y - p = 0, \qquad x + 0y - \bar{p} = 0$$

are parallel by (i), as are also the two lines

$$0x + y - q = 0, \qquad 0x + y - \bar{q} = 0.$$

Let such pairs of lines correspond to geometrically parallel geometric lines. See Figure IV.2.

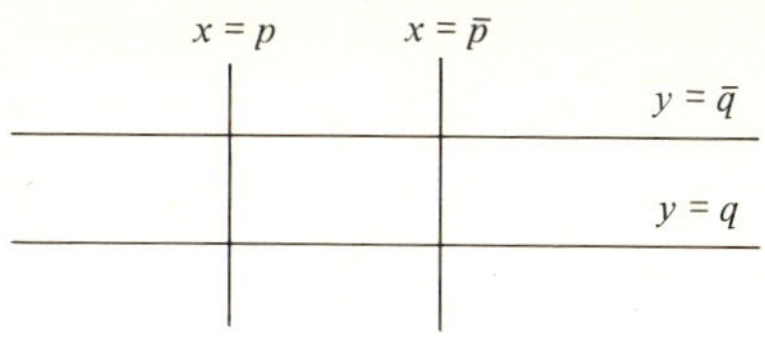

Figure IV.2

Two special lines are

$$0x + y - 0 = 0, \quad x + 0y - 0 = 0,$$

whose geometric counterparts we call the x-axis and the y-axis, respectively. Let P be a geometric point. Draw two geometric lines through P, respectively parallel to the axes. Then these geometric lines correspond to (see (ii) page 273)

$$\begin{aligned} kx + 0y + C &= 0, \quad k \neq 0 \\ 0x + \bar{k}y + \bar{C} &= 0, \quad \bar{k} \neq 0, \end{aligned}$$

which intersect in $(-(C/k), -(\bar{C}/\bar{k})) = (p,q)$, say. Let this number pair correspond to the geometric point P. See Figure IV.3.

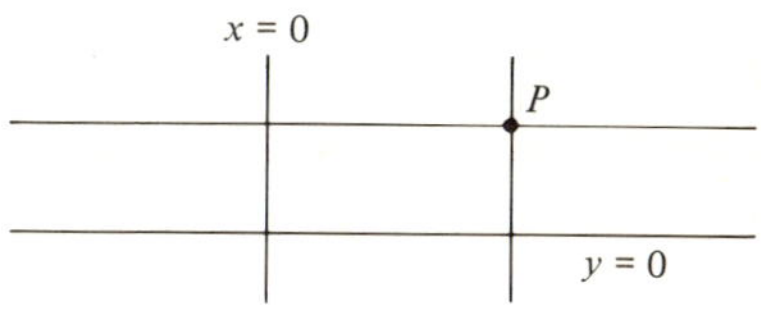

Figure IV.3

Thus we have a one-to-one correspondence between points of the plane π and geometric points of the geometric plane:

$$(p,q) \longleftrightarrow P.$$

Given the geometric point P, we have seen that there corresponds uniquely a number p, and also uniquely a number q. That is, the mappings,

$$\xi: P \longrightarrow p, \qquad \eta: P \longrightarrow q$$

are functions which we call *coordinate variables,*[1]

$$\xi(P) = p, \qquad \eta(P) = q.$$

These coordinate variables do not have inverses. To see why, let

[1] Also see Appendix III.

$$P \longrightarrow (p,q), \qquad \bar{P} \longrightarrow (p,q+1)$$

so that $P \neq \bar{P}$, but $\xi(P) = \xi(\bar{P}) = p$. See Figure IV.4.

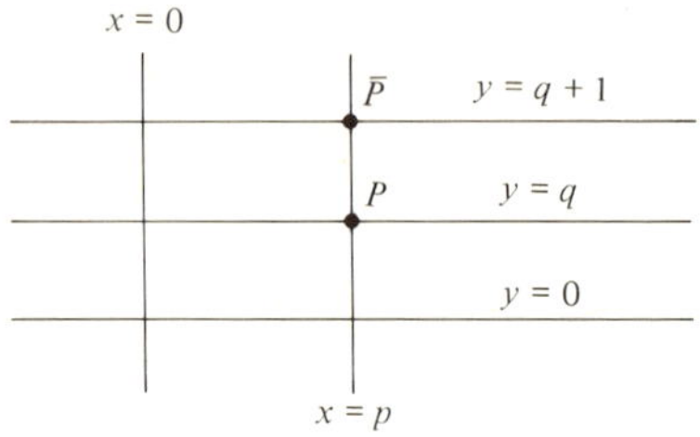

Figure IV.4

To aid in the translation of subsets of π into geometric figures, we define the "length of a segment," $|P\bar{P}|$, where

$$\xi(P) = p, \quad \eta(P) = q$$
$$\xi(\bar{P}) = \bar{p}, \quad \eta(\bar{P}) = \bar{q}$$

to be

$$|P\bar{P}| = \sqrt{(\bar{p} - p)^2 + (\bar{q} - q)^2}.$$

We make the assumption that if P and $\bar{P}$ are on the same axis and if Q and $\bar{Q}$ are on the same axis, then $|P\bar{P}| = |Q\bar{Q}|$ implies that the geometric line segments $P\bar{P}$ and $Q\bar{Q}$ are geometrically congruent. For example, (see Figure IV.5) if

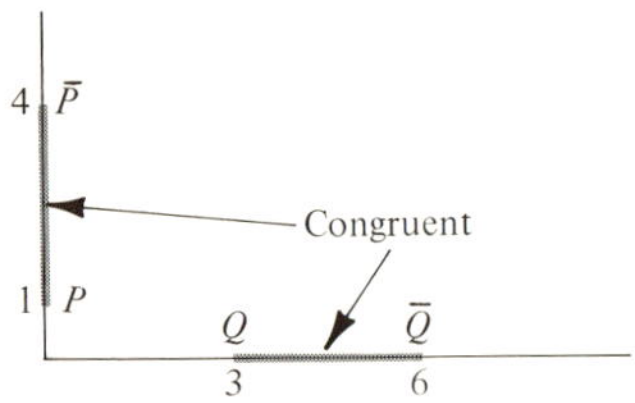

Figure IV.5

$$P \longrightarrow (0,1), \quad \bar{P} \longrightarrow (0,4)$$
$$Q \longrightarrow (3,0), \quad \bar{Q} \longrightarrow (6,0),$$

then

$$|P\bar{P}| = 3, \quad |Q\bar{Q}| = 3,$$

and hence $P\bar{P}$ is geometrically congruent to $Q\bar{Q}$.[2] It can be shown by

[2] To justify this assumption formally, that is, to prove that there exist coordinate variables ξ and η for which this is possible, we would need to have at hand an axiomatic treatment of "geometric" geometry, and we would have to investigate its relation to the plane π and lines Λ.

the geometrical Pythagorean theorem that our assumption leads to the more general result that $|P\bar{P}| = |Q\bar{Q}|$ implies that $P\bar{P}$ is congruent to $Q\bar{Q}$ whether or not these points are on the axes.

To abbreviate the writing we let the symbol $P(p,q)$ stand for

$$\xi(P) = p, \quad \eta(P) = q,$$

where $(p,q) \in \pi$, P is the geometric point corresponding to (p,q) and ξ and η are the coordinate variables. We call the values of these variables at P the *coordinates* of P.

ROTATION, TRANSLATION

In order to reinforce our adoption of the name "line" for the more general set of the type Λ, we define a function R_θ whose domain and range are both the set π. We call this function a *rotation* and describe it by the equations

$$\begin{aligned} R_\theta\colon x' &= x \cos\theta - y \sin\theta \\ y' &= x \sin\theta + y \cos\theta, \quad \theta \text{ is a real number.} \end{aligned}$$

This function R_θ has the inverse R_θ^{-1},

$$\begin{aligned} R_\theta^{-1}\colon x &= x' \cos\theta + y' \sin\theta \\ y &= -x' \sin\theta + y' \cos\theta, \end{aligned}$$

where

(B) $$R_\theta^{-1} = R_{-\theta},$$

and the composition of two relations R_θ, R_ϕ, is again a rotation, namely

(C) $$R_\theta \circ R_\phi = R_{\theta+\phi},$$

as can readily be computed. In particular,

$$R_\theta(k,0) = (k \cos\theta, k \sin\theta), \quad k > 0,$$

and if the number θ is taken to be the radian measure of a geometric angle, the geometric picture is given by Figure IV.6.

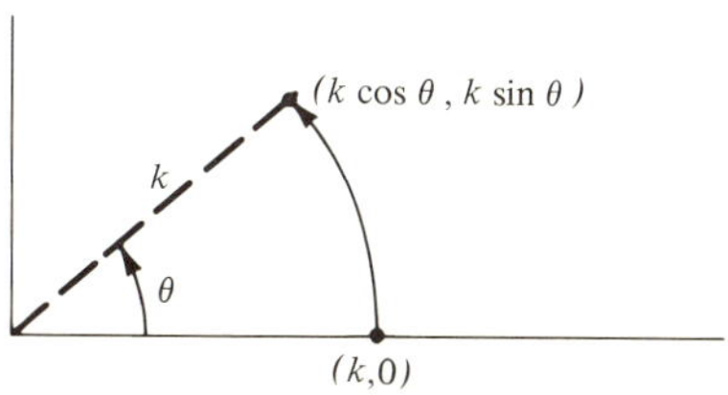

Figure IV.6

Let us interpret $R_\theta(p,q)$ on the diagram. Call $O(0,0)$, which is the intersection of the axes, the *origin,* and let ϕ be the radian measure of the angle from the x-axis to a segment OP. As in Figure IV.7, let P' be a point on the circle with center O and radius $|OP|$ such that θ measures the angle

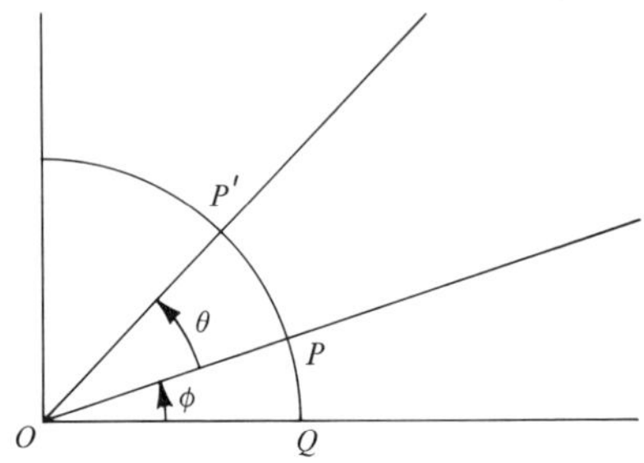

Figure IV.7

from OP to OP'. The point Q is the intersection of the circle with the x-axis. As we saw above,[3]

$$R_\phi(Q) = P$$

and

$$R_{\theta+\phi}(Q) = P'.$$

Therefore,

$$R_{\theta+\phi} \circ R_\phi^{-1}(P) = P'.$$

But from (B) and (C),

$$R_{\theta+\phi} \circ R_\phi^{-1} = R_{\theta+\phi} \circ R_{-\phi} = R_\theta,$$

and so

$$R_\theta(P) = P'.$$

The rotation R_θ carries a point around a circle with center at the origin, through the angle measured by θ.

Probably the most important property of the rotation function is expressed by the formula

$$|R_\theta(P),\, R_\theta(Q)| = |PQ|,$$

that is, the length of the segment formed by the images of two points

[3] The domain of R_ϕ is the set of number pairs, π. However, once a geometric point, such as Q, is identified with a number pair such as $(q,0)$, then either of the point symbols may be used to represent a domain element of R_ϕ. The context determines whether the symbol refers to the analytic or the geometric plane.

under a rotation is the same as the length of the segment formed by the points themselves, as shown in Figure IV.8. A routine computation verifies this. Any subset of π maps into another subset of π by a rotation of its points. As we now see, the geometric image of a subset of π is transformed into a new geometric image in which the distances between pairs of corresponding points is unchanged. Thus R_ϕ "rotates" figures around the origin without changing their shapes.

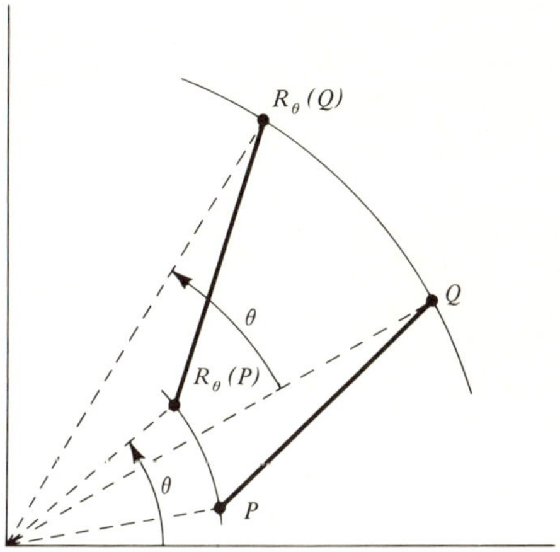

Figure IV.8

Another way to diagram a rotation appears if the points P and $R_\theta(P)$ are superimposed, and new geometric axes are drawn, so that the positions of P and $R_\theta(P)$ relative to these axes are the same as before. Figures IV.9 and IV.10 picture the same rotation R_θ. When some configuration

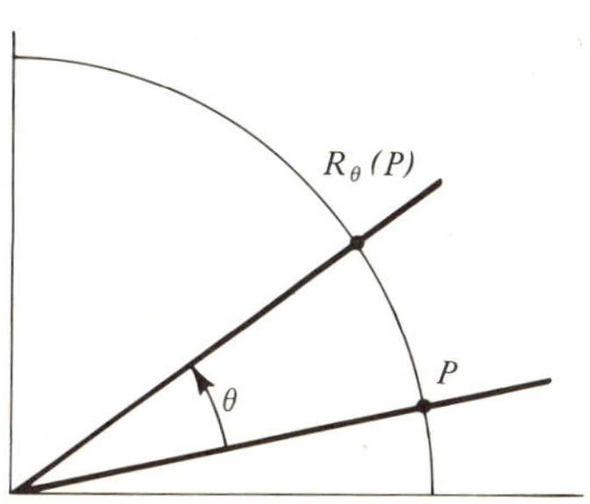

Figure IV.9

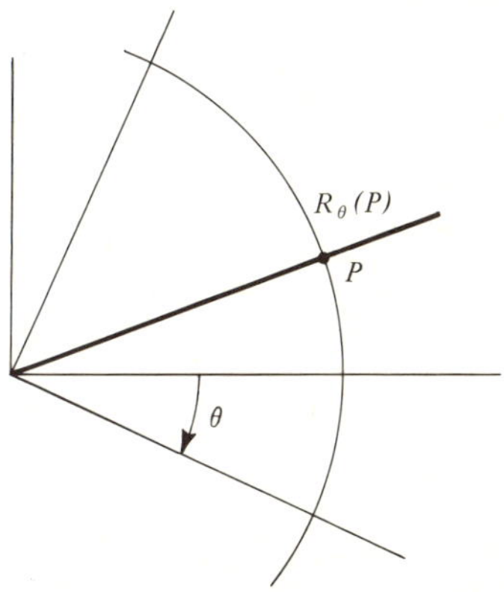

Figure IV.10

(subset of π) is under consideration, the invariance of lengths of segments is emphasized by Figure IV.10. The configuration remains fixed on the page, but is viewed from rotated geometric axes as in Figures IV.11 and IV.12.

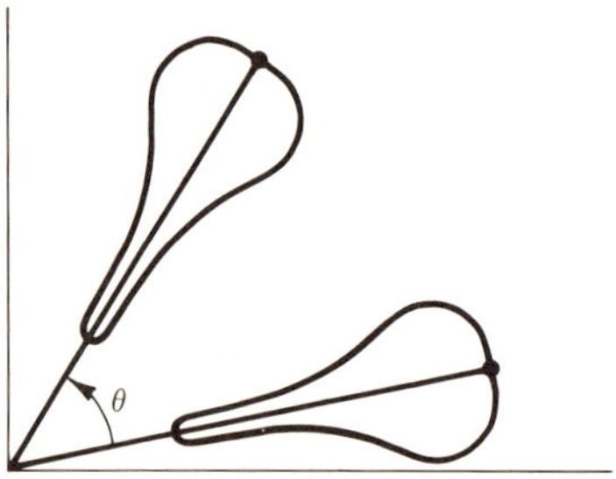

Figure IV.11

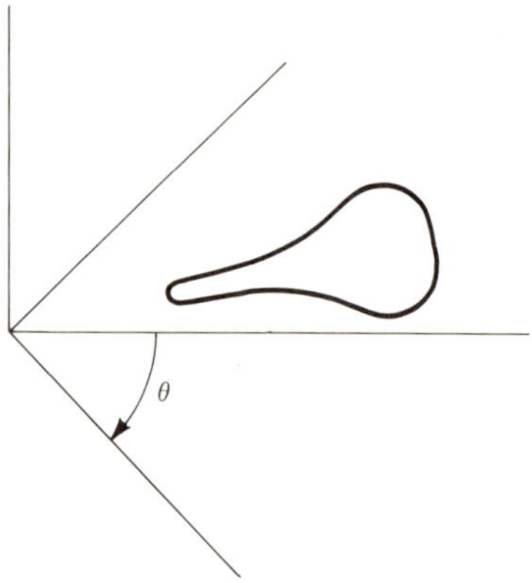

Figure IV.12

We use this interpretation of a rotation to show why we draw a geometric line for the line

$$Ax + By + C = 0 \tag{D}$$

with a straight edge, just as we did for lines such as $x = p$ or $y = q$. Write (D) in the equivalent form

$$\frac{A}{\sqrt{A^2 + B^2}}\, x + \frac{B}{\sqrt{A^2 + B^2}}\, y + \frac{C}{\sqrt{A^2 + B^2}} = 0$$

so we can choose θ with

$$\cos \theta = \frac{A}{\sqrt{A^2 + B^2}}, \qquad \sin \theta = \frac{B}{\sqrt{A^2 + B^2}},$$

and the line (D) is given by

$$x \cos \theta + y \sin \theta - k = 0. \tag{E}$$

If we now rotate this line (E) by $R_{-\theta}$, each point (x,y) in it goes to the point (x',y'), where

$$x = x' \cos \theta - y' \sin \theta$$
$$y = x' \sin \theta + y' \cos \theta.$$

Substituting these formulas in (E) we get

$$x' - k = 0, \tag{F}$$

which we already agreed we would represent by a geometric line. Using the alternative diagram of a rotation we get Figure IV.13, where the geo-

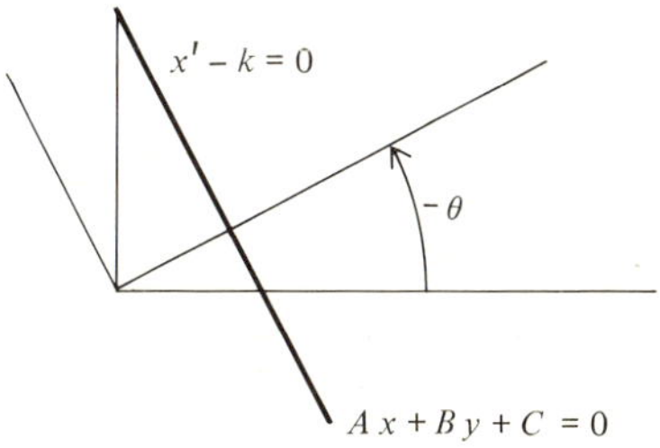

Figure IV.13

metric picture of the line is unchanged, just viewed from different axes.

The mapping of π into π denoted by $T_{(h,k)}$ is given by

$$T_{(h,k)}\colon x' = x + h, \quad y' = y + k, \quad h,k \text{ real numbers,}$$

and is called a *translation*. The reason for this name is clear from either of the geometrical diagrams in Figure IV.14 or IV.15.

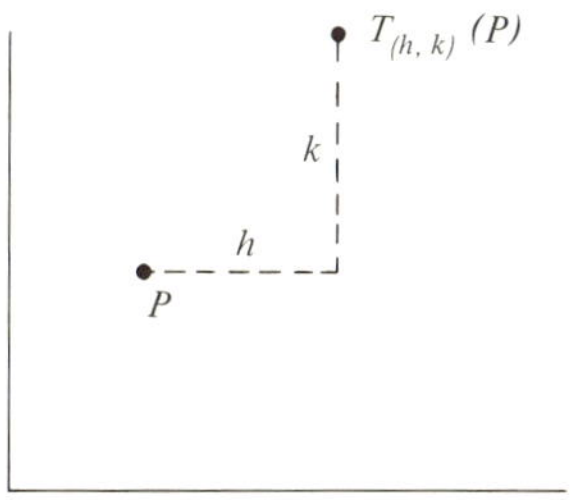

Figure IV.14

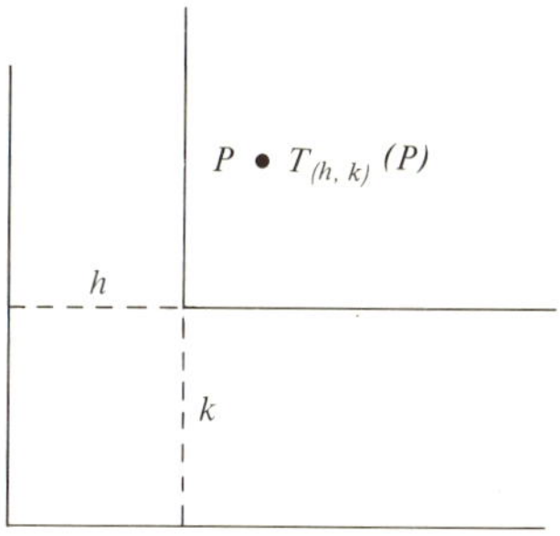

Figure IV.15

Any line can be translated so as to contain the origin. Consider the line

$$Ax + By + C = 0.$$

Under a translation this equation is

$$A(x' - h) + B(y' - k) + C = 0,$$

or

$$Ax' + By' + (C - Ah - Bk) = 0.$$

If $A \neq 0$, take $(h,k) = ((C/A),0)$ to get

$$Ax' + By' = 0,$$

or if $A = 0$, then $B \neq 0$, and $(h,k) = (0,(C/B))$ will do.

Given any line

$$\text{(G)} \qquad Ax + By + C = 0, \quad B \neq 0,$$

it can be rotated so as to become parallel to the x-axis; that is, to be given by

$$\text{(H)} \qquad y' - K = 0.$$

(See also the derivation of Formula (F).) When we rotate (G) by R_θ,

$$A(x' \cos\theta + y' \sin\theta) + B(-x' \sin\theta + y' \cos\theta) + C = 0,$$

or

$$A(\cos\theta - B\sin\theta)x' + (A\sin\theta + B\cos\theta)y' + C = 0,$$

we get (H) when θ is taken to be a solution of

$$\tan\theta = -\frac{A}{B}.$$

The geometric angle θ is the *inclination* of the line, and $\tan\theta$ is the *slope* of the line. Using the usual geometric definition of the tangent function, Figure IV.16 shows that $\tan\theta = b/a$. This is called the *grade* of the line by civil engineers when the line represents a roadbed. The slope is not defined for a line parallel to the y-axis, $B = 0$ (unless one chooses to call this slope ∞).

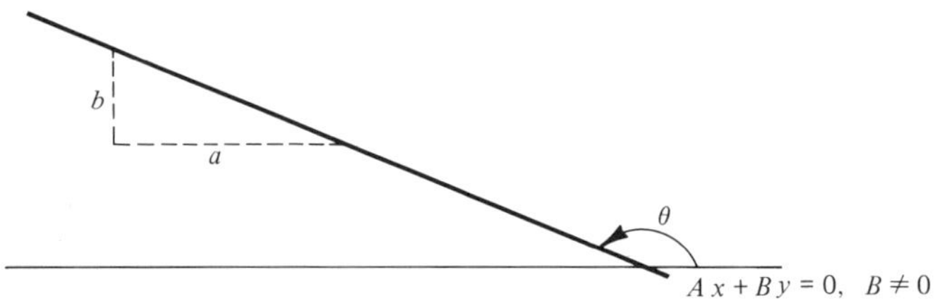

Figure IV.16

The two lines

$$\text{(J}_1\text{)} \qquad Ax + By + C = 0, \quad A \neq 0$$
$$\text{(J}_2\text{)} \qquad \bar{A}x + \bar{B}y + \bar{C} = 0, \quad \bar{B} \neq 0$$

are called *perpendicular* to each other if there is a rotation $R_{\pi/2}$ that carries one into a line parallel with the other.

$$R_{\pi/2}\text{: } x' = -y, \quad y' = x$$

Applied to (J_1),

$$R_{\pi/2}(J_1): -Bx' + Ay' + C = 0,$$

and hence

$$-\frac{\bar{A}}{\bar{B}} = \frac{B}{A}.$$

Then the condition for the mutual perpendicularity of (J_1) and (J_2) takes the form

(K) $$A\bar{A} + B\bar{B} = 0, \quad A\bar{B} \neq 0.$$

If these lines are perpendicular and $A = 0$, it follows that $\bar{B} = 0$ also; therefore, (K) holds without the condition on $A\bar{B}$. If both lines have slopes, the condition (K) can be read "the slope of either line is the negative reciprocal of the slope of the other line." See Figure IV.17.

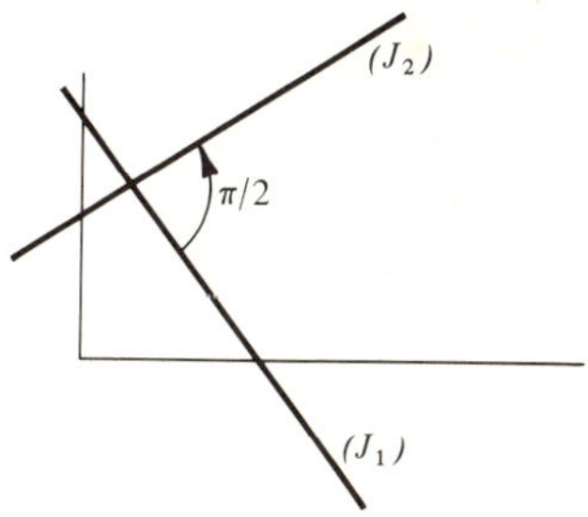

Figure IV.17

CONCLUSION

This completes, in its essentials, our study of subsets of π of the form Λ. Other subsets of π occur commonly, especially those of the form T, where

$$T(A, \ldots, F) = \{(x,y) | Ax^2 + Bxy + Cy^2 + Dx + Ey + F = 0\}.$$

This is a natural extension of the linear equation in x and y to the quadratic equation in x and y. A complete (in its essentials) examination of the subsets T analogous to our study of subsets Λ is long; in fact it is tedious, and we refer you to standard works on analytic geometry for this. In summary, subsets T can be classified as (according to the values of $A, \ldots, F$)

the plane π
the empty set $\varnothing$
a set containing a single point
two lines (*coincident*, parallel, or intersecting)
nondegenerate *conics*.

By the proper choice of rotations and translations, the nondegenerate conics can be reduced to three subclasses. These are

ellipse: $E(a,b) = \left\{(x,y) \,\middle|\, \frac{x^2}{a^2} + \frac{y^2}{b^2} = 1, \quad ab \neq 0\right\}$
(circle, in case $a = b$)

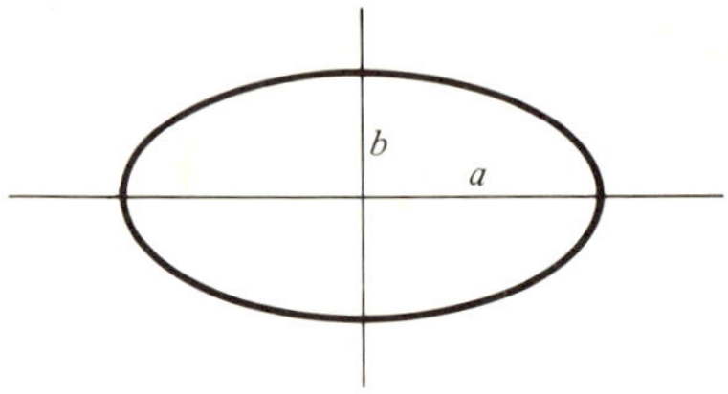

parabola: $P(p) = \{(x,y) \,|\, y^2 = 4px, \quad p > 0\}$

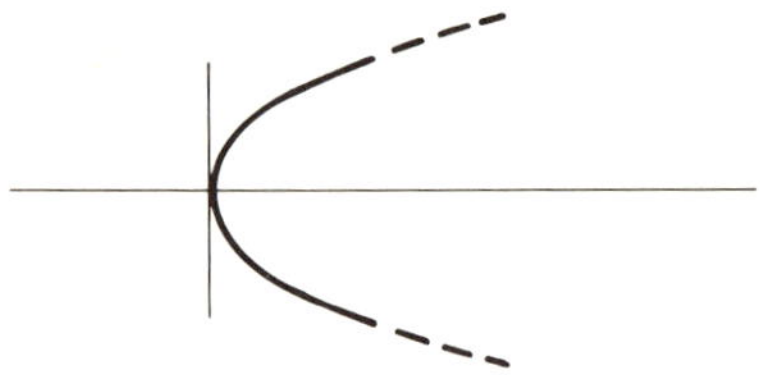

hyperbola: $H(a,b) = \left\{(x,y) \,\middle|\, \frac{x^2}{a^2} - \frac{y^2}{b^2} = 1, \quad ab \neq 0\right\}$
(rectangular hyperbola, in case $a = b$).

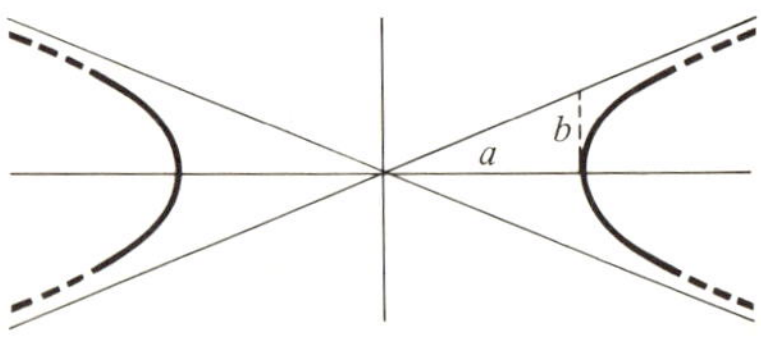

Other subsets of π, of more or less interest depending on one's needs, are legion, and special methods for their examination have been devised. Basically, though, the rotation and translation mappings in π provide the principal tools for studies in analytic euclidean geometry, and we have illustrated here how these tools can be used by applying them to the lines Λ.

INDEX OF SYMBOLS

(Symbols of algebra and notation for elementary functions are not listed.)

INDEX

D

E

F

G

H

I